Technologies and Properties of Modern Utilised Materials

(TPMUM 2012)

IOP Conference Series: Materials Science and Engineering Volume 35

Katowice, Poland
18 May 2012

ISBN: 978-1-63439-846-6
ISSN: 1757-8981

Technologies and Properties of Modern Utilised Materials

(TPMUM 2012)

IOP Conference Series: Materials Science and Engineering Volume 35

Katowice, Poland
18 May 2012

ISBN: 978-1-63439-846-6
ISSN: 1757-8981

IOP Institute of Physics

Printed from e-media with permission by:

Curran Associates, Inc.
57 Morehouse Lane
Red Hook, NY 12571

Printed by Curran Associates, Inc. (2015)

For permission requests, please contact the Institute of Physics
at the address below.

Institute of Physics
Dirac House, Temple Back
Bristol BS1 6BE UK

Phone: 44 1 17 929 7481
Fax: 44 1 17 920 0979

techtracking@iop.org

Additional copies of this publication are available from:

Curran Associates, Inc.
57 Morehouse Lane
Red Hook, NY 12571 USA
Phone: 845-758-0400
Fax: 845-758-2634
Email: curran@proceedings.com
Web: www.proceedings.com

TABLE OF CONTENTS

COATINGS

MICROSTRUCTURE AND CHARACTERIZATION METHODS

DEGRADATION PROCESSES

CERAMICS

COMPOSITES

WELDING

OTHER TOPICS

Peer review statement

All papers published in this volume of *IOP Conference Series: Materials Science and Engineering* have been peer reviewed through processes administered by the proceedings Editors. Reviews were conducted by expert referees to the professional and scientific standards expected of a proceedings journal published by IOP Publishing.

The structure of Fe-Cr-B coatings obtained using selected methods of thermal spraying

K Szymański[1], **M Szpak** [2]
[1]Silesian University of Technology, Faculty of Materials Science and Metallurgy, Krasińskiego 8, 40-019 Katowice, Poland
[2] Energo-Coating, Klejowa 24, 26-600 Radom, Poland

E-mail: krzysztof.szymanski@polsl.pl

Abstract. This paper presents issues connected with the production of protective coatings via thermal spraying. Information about wire coating materials which contain phases from the Fe-Cr-B system is presented. Arc thermal spraying is characterised and its possible applications are determined. The results of the examinations of the structure of coatings obtained by means of various arc spraying systems are discussed. Coatings of this type are used in the machine building and power engineering industries.

1. Introduction

Thermally sprayed coatings are widely used to protect surfaces against abrasive and corrosion wear. They are used, for example, to protect components of power engineering devices and boilers against abrasive wear induced by hard ash particles hitting at different angles and against high operating temperature up to ca. 700 °C, as well as the corrosion caused by the fuel combustion products. Therefore, the coatings that work in such conditions must have high hardness and required thickness, good adhesion to the substrate, low porosity and a complex phase composition which will guarantee a high corrosive resistance to compounds containing sulphur, chlorine and water vapour in oxygen depletion [1-5].

Among a number of methods of thermal spraying, the ones that are practically used for protecting boiler components are the flame method, arc spraying, plasma spraying and high velocity oxygen fuel spraying (HVOF). The flame methods and plasma spraying are relatively rarely used due to the complexity of the process (flame methods with fusing) or relatively high costs (plasma spraying) [6-8].

The most frequently used methods of thermal spraying used for protecting boilers are arc and high velocity methods. The HVOF thermal spraying allows obtaining highest quality coatings of carbide materials. They are characterised by: high hardness (up to 1600 HV), tightness (porosity <2 %) and good adhesion to the substrate (>80 MPa). However, they are relatively expensive. The characteristics of HVOF methods is a broad issue which has been addressed in many papers [9-13].

On the other hand, the biggest advantage of the arc method is the possibility to effectively cover huge areas, which translates into relatively low production costs of the coating. The development of the method and good functional properties of coatings result from the application of advanced

[1] To whom any correspondence should be addressed.

composite materials with a pre-designed phase composition, with a metallic matrix resistant to corrosion and hard reinforcement phases. Properly produced coatings have a banded structure of low porosity (ca. 2 %), a high level of thickness (0.5-1 mm) and varied hardness, from ca. 350 (IN635) to 1000-1200 HV (e.g. W-685 and CE-595). The different properties of coatings may be obtained by using materials with diverse chemical and phase compositions, and optimized parameters of coating production. Examples of used wire materials are presented in Table1.

Table 1. Examples of wire materials which contain phases from the Fe-Cr-B system.

No.	Material name	Approximate chemical composition of the wires
1.	FMI - 3	Fe-Al-Cr-B
2.	W-685	Fe-Cr-B
3.	M 5	Fe-Cr- C
4.	SF 27/	Fe-Ni-Cr-Si-B
5	FMI - 2	Fe-Al- Cr-B
6.	CE- 595	Fe-Cr-Si-B

2. Aim and scope of the research

The arc-sprayed coating materials presented in Table 1 are distinguished by a number of advantageous functional properties. However, the best coatings are obtained by using very good thermal spraying devices and correctly selecting the parameters of coating production. Therefore, the essential purpose of the research presented in this paper was to examine the influence of the type of the device on the structure of coatings obtained from the coating material which had been adopted as a model. Chemical composition of the sprayed material is presented in Table 2.

Table 2. Approximate chemical composition of the material used in the tests.

Percentage content of chemical elements in the wire							
Fe	**C**	**Si**	**Mn**	**Cr**	**Mo**	**W**	**B**
balance	1.05	1	0.9	22	1.2	1.25	3.4

Coatings were sprayed onto plates made of the P235 steel with surface roughness R_a>10µm, previously cleaned by abrasive blasting to the SA3 class. Three devices were used in the process of thermal spraying: Smart ARC (Sulzer Metco), AT 400 (Flame Spray Technologies) and OSU 300A LD/U-2 (OSU Hessler). The coatings were sprayed at varied process parameters which were selected individually for each device - their range is presented in Table 3.

Table 3. Devices and the range of spraying parameters.

Device	Spraying parameters			
	Voltage (V)	**Amperage (A)**	**Distance (cm)**	**Other**
Smart ARC	32	280	10-15	HV nozzle
AT 400	36	320	10-15	standard nozzle
OSU	25	180	10-15	open system

After the process of spraying, the macrostructure of the coatings was examined – researchers looked for exfoliations, discontinuities and other superficial faults. Positively evaluated samples with the required coating thickness were submitted for metallographic examination. Microsections were observed on an Olympus GX51 light microscope and a Hitachi S3400N scanning microscope with an analysis of chemical composition. Porosity was calculated using the Metilo application which automatically conducts a quantitative image analysis.

3. Research results

The examined coatings have demonstrated a correct macrostructure – homogeneous with no surface defects, had the prescribed thickness. The coating sprayed using the OSU system with an open nozzle had visibly higher roughness. Examples of the structure of the obtained coatings are presented in Figures 1-3.

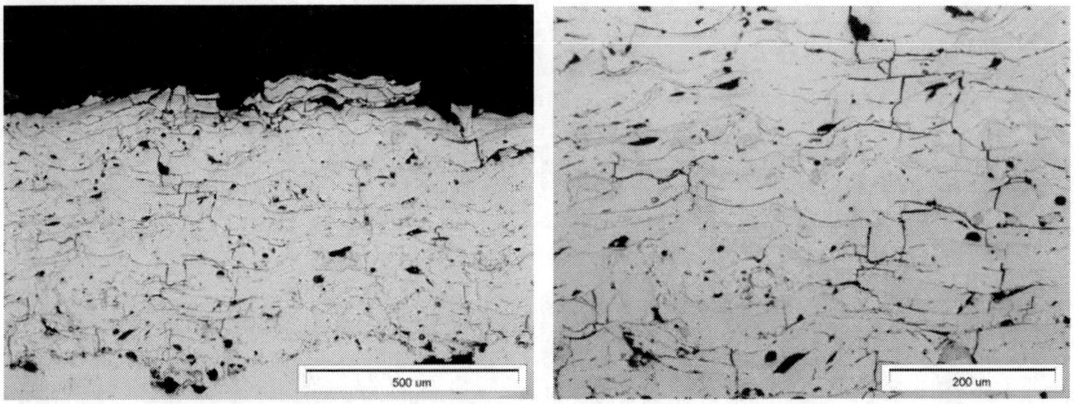

Figure 1. The structure of coatings sprayed with the AT400 device.

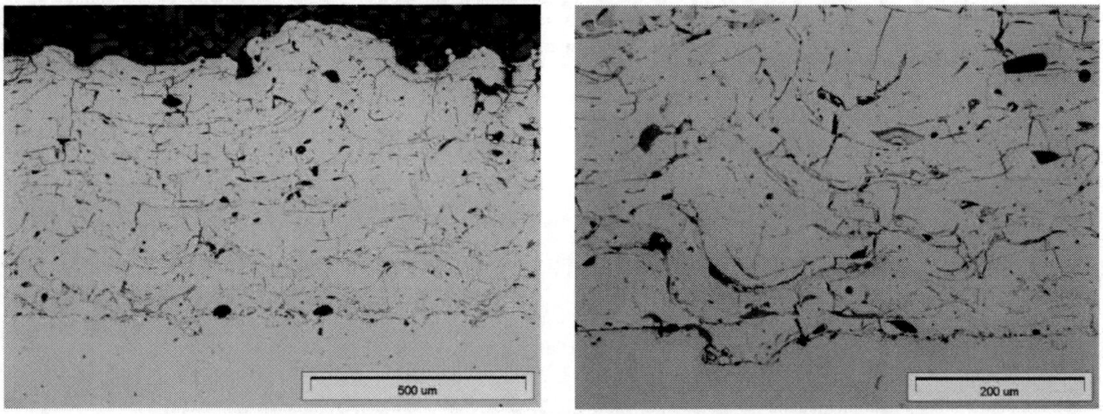

Figure 2. The structure of coatings sprayed with the Smart ARC device.

The coating obtained using the OSU system with an open nozzle exhibited an unsatisfactory structure - discontinuity in the coating and was excluded from further tests. The structures of coatings obtained with the use of Smart ARC and AT 400 after an adjustment of spraying parameters had comparable hardness indices and void ratios (Table 4).

The coating obtained using the Smart ARC system was examined on a scanning microscope (Fig. 4) and its chemical composition was determined with the application of EDX.

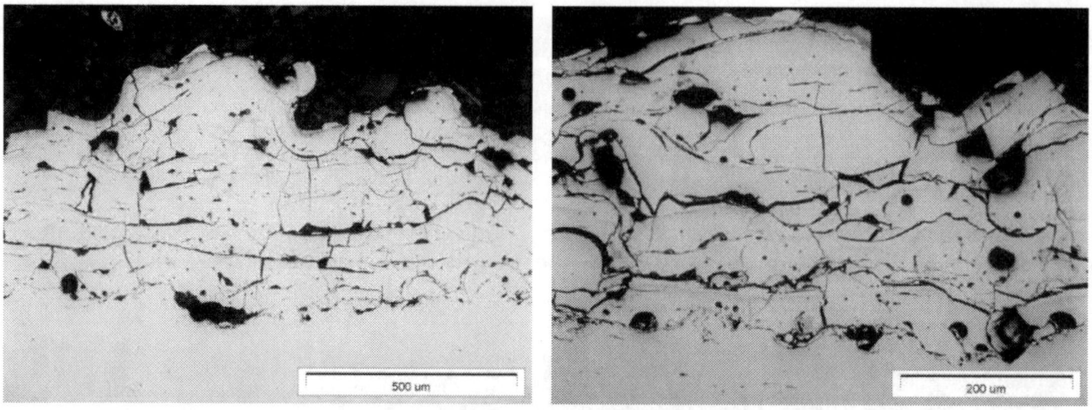

Figure 3. The structure of coatings sprayed with the OSU device.

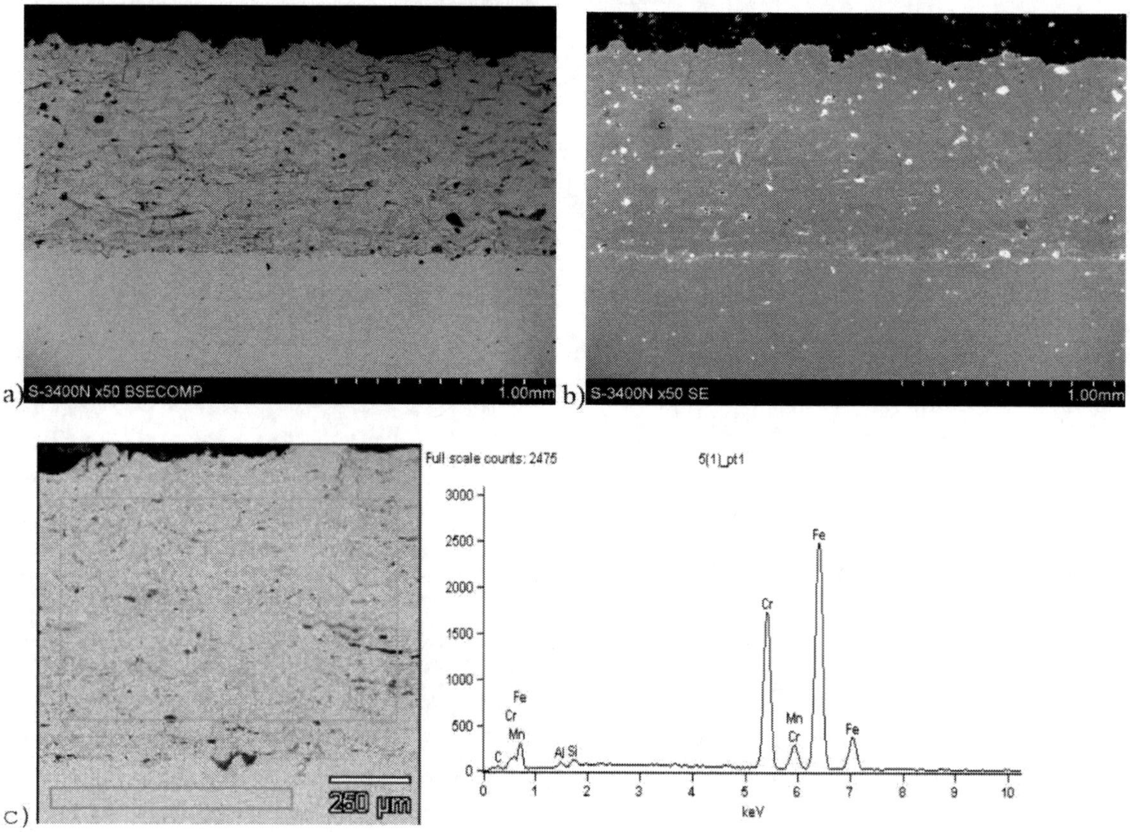

Figure 4. The structure of the coating sprayed using Smart ARC: a) SEM, b) BSE, c) EDX.

The structure (Figs 2, 4) and the results of the EDX analysis of the chemical composition indicate high homogeneity and compactness of the coating which adheres to the substrate very well. The obtained microsections were used for tests of porosity. For this purpose, the automatic image analysis methods (the Metilo software) were used. Hardness of the coatings was also determined. The results are presented in Table 5.

Table 4. Properties of the obtained coatings.

Device	Coating parameters		
	Porosity (%)	**Hardness (μHV)**	**other**
Smart ARC	1.8	1200	-
AT 400	2.7	1100	-
OSU	-	-	fractures

Table 5. Chemical composition of the obtained coatings – atomic %.

Area	Al	Si	Cr	Mn	Fe
pt1	1.3	2.0	29.2	1.0	66.5
pt2		0.7		0.5	98.7

4. Conclusions

Coatings obtained by means of arc thermal spraying are distinguished by a correct structure (low porosity, good adhesion to the substrate) and very high hardness which is comparable to carbide coatings applied through high velocity oxygen fuel spraying. It is possible to produce relatively thick coatings (up to ca. 500μm) which are characterised by high resistance to abrasive and corrosion wear [9, 14]. Therefore, it is possible to use them for the production of coatings which protect different components against erosion and corrosion wear, while also operating at elevated temperatures. Such coatings are frequently used to protect the walls of combustion chambers in power boilers. Examples of such applications are shown in Figure 5.

Figure 5. Arc spraying of coatings made of Fe-Cr-B material on the components of power boilers.

Unfortunately, coatings of this type tend to be susceptive to the formation of single grain fractures vertical to the surface or fractures between grains. This phenomenon is characteristic for the examined coating material, however, a proper selection of spraying parameters significantly reduces this defect. A correction of the chemical composition of the coating material provides better possibilities of avoiding fractures in the material layers. For example, an addition of Al improves, among others, plastic properties of the coating [15-17]. The application of additional sealing coatings [18] is advantageous for components working at high temperatures and in an aggressive environment.

Systems which are designed to spray "hard" wires should be used to spray coatings of the Fe-Cr-B type. Coatings produced using the OSU system with an open nozzle do not meet the strict requirements set upon the coatings to be applied in the power engineering industry.

5. References

[1] Formanek B, Szymański K, Hernas A et. all 2002 Erosion damage and protective coatings for water walls of CFB boilers, *AMME*.
[2] Hoop P J and Allen C 1999 *Wear* **233–235** 22.
[3] Stringer J 1995 *Wear* **186-187** 312.
[4] Wang B Q 1996, Erosion –corrosion of thermal sprayed coatings in FBC boilers, *Wear* **188-189** 24.
[5] Formanek B, Szymański K, Hernas A, Letsko A and Ilyuszenko A 2011, Thermally sprayed composite coatings for corrosion protection of power energy boilers, *2 International Symposium «Surface Engineering. New Powder Composition Materials. Welding»* March 23-25, Minsk, Belarus.
[6] Higuera Hidalgo V, Belzunce Varela J, Ceries Mendez A and Poreda Martinez S 2001 *Wear* **247** 214.
[7] Higuera Hidalgo V, Belzunce Varela J, Ceries Mendez A and Poreda Martinez 2001 *Tribology Int.* **34** 161.
[8] Buta Singh Sidhu and Puri SPrakash D 2004 *Mater. Sci. Eng.* A **368** 148.
[9] Formanek B, Szymański K and HernAS A 2011 *Material Engineering* No 4 182 399.
[10] Wang B Q and Luer K 1994 *Wear* **174** 142.
[11] Verstak A, Wang B Q, Baranovski V and Bielaiev A 1998 *Corrosion 98*, paper no 193, NACE, Houston.
[12] Formanek B 1997 *Biuletyn Instytutu Spawalnictwa w Gliwicach* (The Bulletin of Welding Institute in Gliwice) No 5.
[13] Szymański K, Formanek B and Kuczowitz B 2009 *Binding* **3** 22.
[14] Formanek B and Szymański K 2006 *Problems of corrosion and corrosion protection of structural materials Special Issue* **5** 198.
[15] Pokhmurskii V, Student M, Pokhmurska V H, Wielage B and Wank A 2005 Influence of solid lubricant addition on tribological properies of coatings arc sprayed with cored wires based on Fe-Cr-B-Al and Fe-Cr-C-Al systems, *Internationale Tagung Thermisches Spritzen*, Wroclaw, 26-28.09.2005, ISBN 83-87982-36-9, 122.
[16] Tillmann W, Vogli E and Ferreira 2008, *Proc. International Thermal Spraying Conference*. Mastritcht, 2008.
[17] Pokhmurska H, Wielage B, Grund T, Schuberth S, Pokhmurskii V, Student M and Karpenko G V 2009 Gas-Abrasive Resistance of Arc Sprayed Dispersion-Hardened Coatings Obtained from Iron-Based Cored Wires, *International Thermal Spray Conference* 2009; Las Vegas USA.
[18] Formanek B, Szymański K and Szczucka-Lasota B 2005, New generation of protective coatings intended for the power industry, *Journal of Materials Processing Technology* **164-165** 21.

Liquid metal embrittlement of steel with galvanized coatings

J Mendala
Department of Materials Technology, Silesian University of Technology, Krasinskiego 8, 40-019 Katowice, Poland

E-mail: jacek.mendala@polsl.pl

Abstract. Article presents the state of knowledge relating occurrences of liquid metals embrittlement. The results of experimental investigations of the LME phenomenon, reasons of its formation and influence of different parameters are described. Selected ideas of the applied research methods accessible in different works are presented. Samples made of C70D steel with tensions stretching (400-800 MPa) at different values and kinds of loading during the hot dip metallization were investigated. Coating processes in two bath: zinc and zinc with 2 % tin addition were made. The processes of hot dip metallization were done at 450 °C temperature and immersion time 180 s. Coated samples were investigated by light microscope to specify possibility appear of LME effect.

1. Introduction

Commonly used constructional steels are easily affected by corrosion. Various methods of protecting their surfaces are used to prevent this. Deposition of protective metal layers through dip coating is the basic method of protecting steel elements against corrosion [1]. During this process, an element is immersed in molten metal and thus, a direct contact occurs between the solid metal and liquid metal, which usually leads to changes of mechanical properties and in extreme cases, even to premature destruction of the element during its further operation. In 1928, Peter Rehbinder [2] initiated research on the unknown phenomenon of cracking of steel structures previously subjected to a hot-dip metallizing process. The effect of a surface-active medium on the mechanical properties of other materials, called the Rehbinder effect, resulting from a drop in surface voltage, has been examined for many types of solid bodies. The phenomenon of environment sensitive mechanical behaviour (ESMB) of solid bodies is very diverse in terms of its occurrence and mechanisms. The most extreme form is the decrease in strength and crushing of metals and alloys when in contact with liquid metal. It is this form of ESMB that has been rendered responsible for a drastic decrease in ductility and brittle cracking of metal (in a solid state) which was previously in direct contact with a liquid metal. This phenomenon is referred to in the literature as liquid metal embrittlement (LME) or liquid metal assisted cracking (LMAC) (figure 1) [3-5,9].

In the 60s of last century, this problem started attracting much attention from scientists and today, extensive studies of the causes of LME are conducted. The interest in this phenomenon has been additionally necessitated by the growing demands of clients who require manufacturers to use engineering materials that will fulfil the expected utility functions in a reliable manner, for as long as possible.

Finding an explanation for the mechanisms of the phenomenon allows finding solutions to certain practical problems, such as protection against destruction caused by contact of a stressed material with molten metal, or using the phenomenon to increase the efficacy of certain actions, such as crushing or reaming of very hard metals.

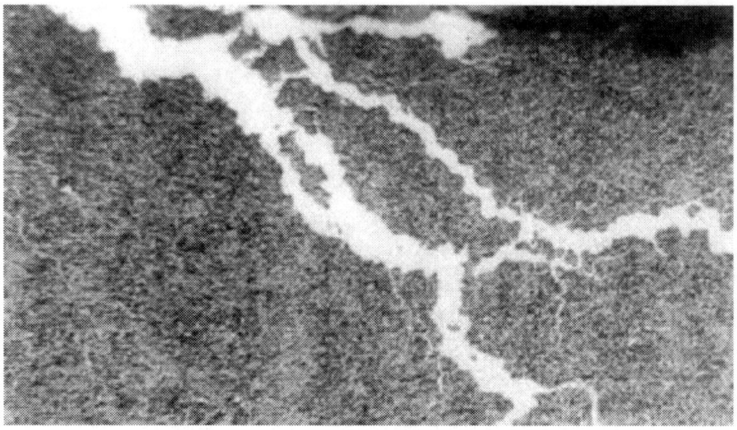

Figure 1. LME of steel dipped in liquid tin at temperature 266 °C [3].

The risk of embrittlement induced by liquid metals has been confirmed, however, it is still difficult to describe it because the knowledge of the phenomenon mechanisms is limited and the number of parameters that seem to be responsible for LME is high. The results of experimental tests of LME described in the literature [3,4] allowed presenting selected mechanisms which broaden the knowledge about this phenomenon. The authors of this study have focused on a detailed examination of literature discussing the LME phenomenon, its causes and the description of the test stands and research methodologies used.

The paper presents a part of research on the influence of the chemical composition of liquid metal, in this case of tin addition in a zinc bath, on the possibility of LME occurrence.

2. LME and the causes of its occurrence

LME can be defined as brittleness or a loss of ductility of a usually plastic material in the presence of liquid metal. Some authors add to this definition the condition that external load or internal residual stresses are necessary for LME to occur [4].

The results of conducted experiments have allowed the characterisation of the LME phenomenon. First experiments consisted in examining the influence of a few parameters which were changed in an inert medium and in a liquid metal environment. A number of parameters which may influence the above-mentioned phenomenon were examined and described.

The main parameters that have a profound impact on the occurrence and course of embrittlement under the influence of liquid metal are:

- the fracture propagation rate – a characteristic feature of LME is a very high speed of fracture propagation compared to the speed of propagation in the air or in a vacuum. A fracture occurs after exceeding a certain threshold value of stress. The speed of propagation is mainly caused by the fact that this phenomenon occurs with various intensity for the pair liquid metal – solid metal [4,5];
- deformation during fracture – in almost every case described in the literature, a tensile test conducted in liquid metal demonstrates that stress during fracture is reduced, compared to the same tests conducted in the air and in a vacuum;
- stress curves – some discrepancies become apparent while comparing the stress curves described in the literature [6,7]. Westwood and Stoloff believe that until a fracture occurs, the stress of solid metal is the same as in the absence of immersion. Popovich [8] claims that

initial development of plastic flow takes place before a fracture. In both cases, elastic properties remain unchanged in the presence of liquid metal;

- the chemical composition of liquid metal – liquid metal influences the course of the phenomenon at the end of a crack by decreasing the critical intensity of cracking stresses and changing its micromechanisms. If we consider the effect of additives in liquid metal, they may increase or decrease brittleness. The composition of liquid metal also modifies a number of physicochemical parameters of the solid metal – liquid metal system (surface energy and intersolubility), which is assumed to influence the LME mechanism [3,4,9];
- the chemical composition of solid metal – the effect of alloying additions in various materials on LME was the object of studies described in the literature. It was determined that some additions increase the LME effect;
- the microstructure of solid metal – the hardness and resistance to deformation of solid metal during loading may affect its susceptibility to LME. An increase in the intensity of the phenomenon may be compensated by cold plastic working, because it reduces the grain size. It was determined [4] that LME intensifies when the grain size increases. There is a linear dependence between the tensile force and grain diameter ($d^{1/2}$). Furthermore, the brittle to plastic transition increases as the temperature grows, due to the grain growth, and is a linear dependence of temperature and log(d);
- wettability – wetting is considered a prerequisite for the occurrence of LME. If contact between the solid and liquid metal is very short or it is interrupted due to the presence of an oxide film, LME does not occur [4,5];
- surface energy and grain boundaries – surface energy γ_{SC} and energy at grain boundaries γ_Z are the most important physicochemical parameters of the solid metal – liquid metal system in the LME model described by Clegg [5].

Much progress has been made in the field of study of the effect of external parameters. Namely:

- temperature – LME occurs in a certain range of temperature called “a drop in ductility”. This property is usually considered to be a dependence of both the solid and liquid metal compositions, as well as of the solid metal structure and external conditions. The lower boundary of temperature required for the occurrence of LME depends on the properties of a given liquid metal. Apart form that, the temperature has an effect on LME kinetics;
- the deformation rate – the dependence between the deformation rate and the speed of crack development is almost linear in certain solid metal – liquid metal systems [4]. During slow deformation in conditions conducive to intensive relaxation of stresses, plastic flow is weakened and can hinder the LME process. A reduction of the deformation rate at an increased temperature leads to a decrease or decay of the range in which embrittlement occurs;
- stress – where solid material is subject to constant stresses, a liquid metal will penetrate the base material, thereby inducing cracks very similar to corrosion cracks. The speed of this process changes depending on the temperature and the type of metal used. The closer the melting point is, the faster the reaction proceeds. The damage caused in this way is catastrophic and often confused with hydrogen embrittlement damage.

The LME phenomenon was analysed and described in the literature by using various research methods which differ from one another mainly in terms of sample profiles and the method of applying stress.

3. Test stand

The basic idea was to create a test stand that would facilitate the induction of tensile stresses in a sample immersed in liquid metal. The sample was to be obtained from steel with a diameter no bigger than 3.5 mm. A standard laboratory furnace was used to produce coatings in a batch hot-dip metallizing. Such assumptions determined the requirements for the test stand. Figure 2 presents the conception of the test stand design.

The LME test stand is comprised of two principal modules (figure 3):

- a furnace, PET 22/45, equipped with a crucible made of nitrided silicon carbide with a capacity of 3.2 dm^3, and a system for measuring and adjusting the furnace temperature to an accuracy of ± 2°C. The upper plate of the furnace is the base for the installation of the second module of the stand, which module is used to induce tensile stresses in a sample dipped in the metallizing bath;
- the device is equipped with a mechano-electrical weight transducer, PW1-S-2.5 type (maximum load – 2.5 Mg) which converts compressive force into an electric signal proportional to the force. The conversion is conducted using the principle of resistance strain gauge measurements. The weight transducer consists of a set of extensometers connected into a full bridge system and stuck to the elastic component of the transducer. The load is applied to the sample indirectly through a screw gear acting on the elastic element of the transducer which ends with a spherical head. This component is closed in a casing that cannot be disassembled. Data from the weight transducer is sent to the weight terminal.

The terminal is equipped with an electronic display which shows the current or maximum values of load acting on the sample. The terminal is equipped with an RS 232 serial port which enables computer registration and recording of the course of tests (values of load applied to the sample with 5 measurements per second).

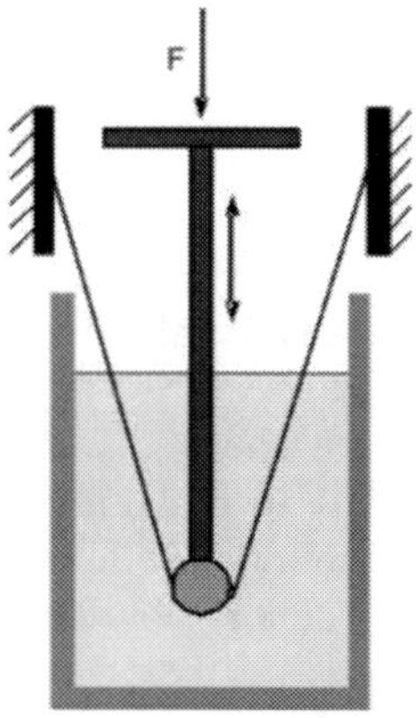

Figure 2. Conception of standing – loading is set in the middle of the sample

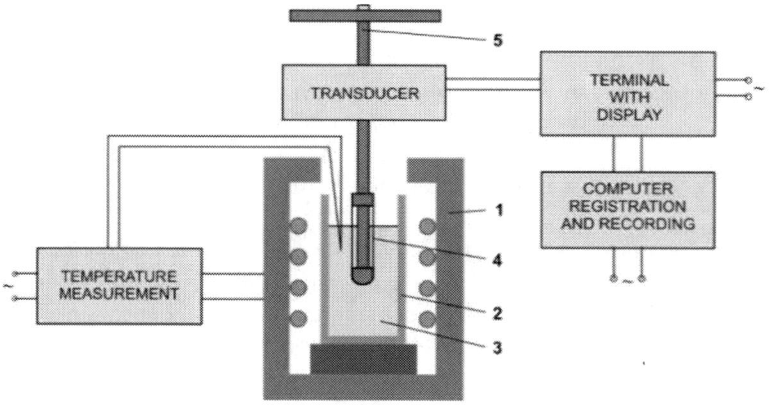

Figure 3. Diagram of standing 1) furnace PET 22/45, 2) ceramic crucible, 3) liquid metal, 4) sample, 5) gear

4. Research methodology

Samples made of 1.0615 (C70D) with a Ø 3 mm diameter were used for tests at variable loads. In order to determine basic strength properties of the material intended for tests, a static tensile test was conducted on the Instron 4469 testing machine, during which strength values were determined.

The tests were conducted in two metallizing baths: a zinc bath and a zinc bath with a 2 wt. % tin addition. To prepare the metallizing bath, zinc and tin of 99.99 % purity were applied, as well as a technological addition of AR0 aluminium in the amount of 0.1 wt. %.

Before starting the metallizing process, the samples were subjected to abrasive blasting and next, to chemical surface preparation in order to enable the best possible contact between the surface of a sample and the bath. The process consisted of the following operations: degreasing in trichloroethylene, etching in a 18 % HCl solution, flux treatment using "Tegoflux" and drying in a chamber drier at the air stream temperature of 160 °C.

The hot-dip metallizing process of the prepared samples was conducted using the batch method (figure 4). It was assumed that the bath temperature will be 450 °C, with metallization time of

180 seconds. Measurement of the liquid metal temperature was taken directly in the metallizing bath using a Ni–CrNi thermocouple placed in a ceramic casing. Prior to the submersion of the sample, initial load corresponding to the stress of 400 MPa was applied. After immersing, additional load was applied to the samples using two different methods:

- one-time application of load to a sample up to the prescribed value immediately after dipping (method I),
- applying additional load to the sample up to the prescribed value immediately after dipping and maintaining this value during metallization (method II).

Each time, just before the sample was pulled out of the bath, the load applied to the samples was reduced to the value corresponding to the stress of about 150 MPa. After the process had been completed, the samples were cooled in the air. Computer registration of the stress started the moment initial load was applied before the submersion and ended when the stress in the sample became stable (about 180-420 seconds) after the lifting. Tests for both methods were conducted at loads corresponding to stresses in the range of 400 to 800 MPa. The plan of the experiment is presented in table 1.

Figure 4. Standing in horizontal position – sample installing (a) and after dipping in bath (b).

Table 1. The plan of the experiment.

Bath	Dipping time (s)	Initial stress (MPa)	Stress during metallizing (MPa)	Stress relieve (MPa)	Kind of load
	180	400	400	150	
	180	400	500	150	method I
Zn	180	400	600	150	and
	180	400	700	150	method II
	180	400	800	150	
	180	400	400	150	
	180	400	500	150	method I
Zn + 2 wt. % Sn	180	400	600	150	and
	180	400	700	150	method II
	180	400	800	150	

Computer registration of tests results allowed the preparation of diagrams illustrating the dependence of stress on time. The results obtained were used in further analysis and observations with a light microscope.

5. Tests results

While analyzing the curves obtained in method I (figures 5 a and 6 a), it can be noted that in each case, there is a sudden decrease in stress from the initial load value to ca. 150 MPa during the first 15 seconds, which corresponds to the moment the sample is immersed in the metallizing bath. Next, the stress increases due to the application of specific load, after which it decreases proportionally to the increase of the load value. After unloading the samples, which is connected with pulling them up from the metallizing bath, (i.e. after 180 s) and their cooling down, one can observe an increase of stress up to a value independent from the applied initial load and specific load.

The course of curves obtained in the case of method II (figures 5 b and 6 b) is analogous to those obtained in method I, with the only difference that the stress does not decrease after the application of specific load, but is maintained at a certain level. The curve in the diagram for the sample that underwent failure before the metallization process was completed is an exception to this. The tests for a sample metallized in a Zn + 2 wt. % Sn bath subjected to a load corresponding to the stress of 800 MPa (method II) were repeated in order to verify the results.

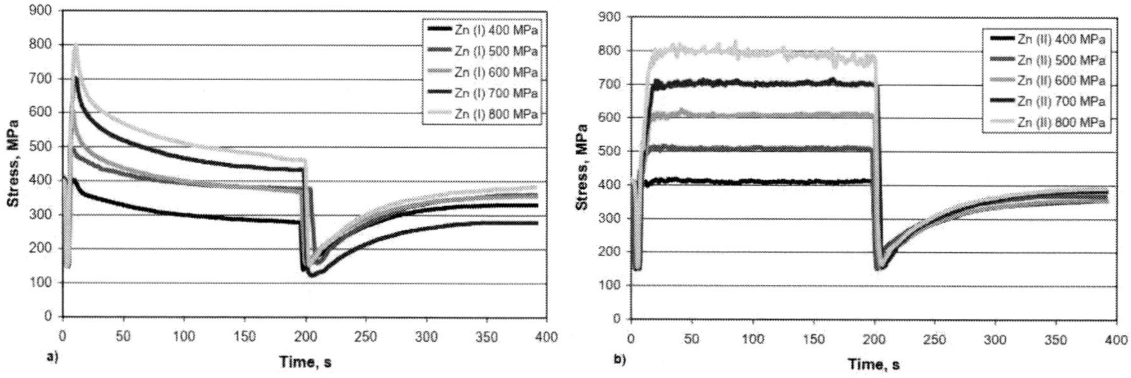

Figure 5. Stress vs. time diagram obtained during metallization of 1.0615 (C70D) steel samples in Zn bath; load corresponding to the stress 400-800 MPa by: method I (a) and method II (b)

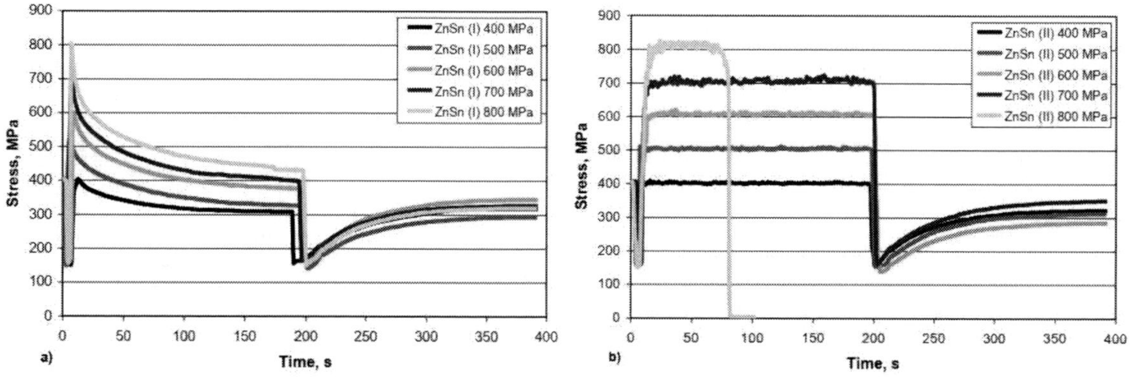

Figure 6. Stress vs. time diagram obtained during metallization of 1.0615 (C70D) steel samples in Zn + 2 wt. % Sn bath; load corresponding to the stress 400-800 MPa by: method I (a) and method II (b)

Based on the above-mentioned results, it can be affirmed that the tin addition introduced into the metallizing bath may decrease the strength properties of the product being coated, meaning that in the metallized component, where the described stress value is present, cracking may occur or the component may be completely destroyed.

All obtained coated components underwent metallographic tests using a light microscope in order to assess the possibility of the LME occurrence. The structure of components with coatings, obtained in both tested bath and subjected to stresses by means of method I, did not show the occurrence of fractures in the substrate material or the coating area. The situation was similar in the case where stress was applied according to method II during the metallization process in a zinc bath, where no fractures were found (figures 7a and b). On the other hand, in the case of applying stress during metallization in a zinc bath with an addition of tin (figures 7c and d), for higher stress values (600-800 MPa) small fractures were detected in the coating which were vertical to the substrate and the sample was ruptured (800 MPa). Furthermore, a detailed microscopic analysis showed the presence of fractures characteristic for the LME phenomenon described in the literature for a sample metallized in a Zn + 2 wt. % bath subject to stress of 800 MPa (figure 8).

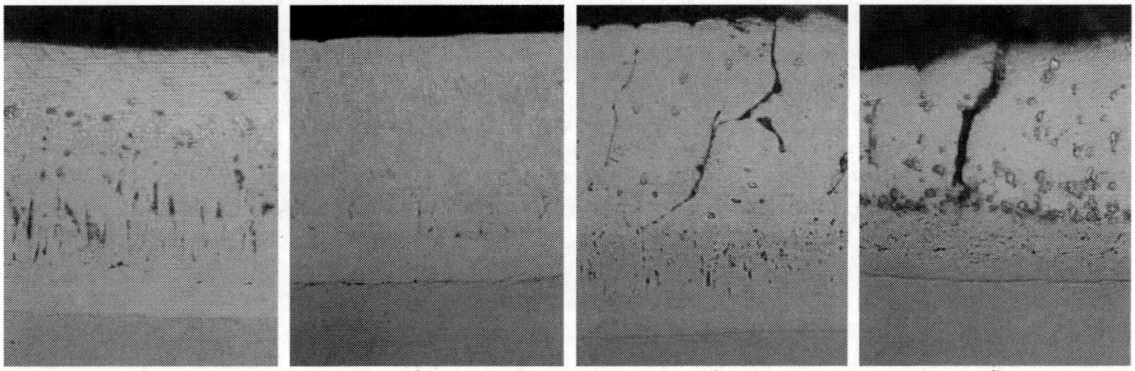

a) b) c) d)

Figure 7. Cross section of samples with coatings subjected to stress obtained in: Zn bath – 600 MPa (a), 700 MPa (b); Zn + 2 wt. % Sn bath – 600 MPa (c), 700 MPa (d).

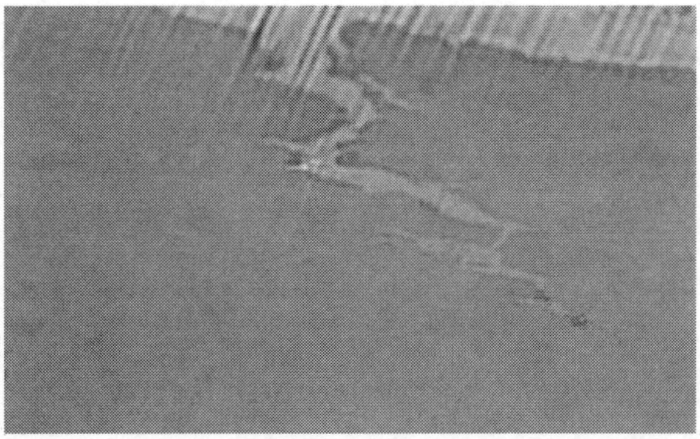

Figure 8. Cross section of sample with coating obtained in Zn + 2 wt. % Sn bath subjected to stress of 800 MPa with fractures characteristic for the LME phenomenon.

6. Conclusions

Based on the assessment of the current state of knowledge, it can be concluded that the LME phenomenon is responsible for the formation of brittle cracks and a sharp decrease in the strength of solid metal during contact with liquid metal. It was noticed that many properties of solid and liquid metal, i.e. the structure of solid metal, the chemical composition of both solid and liquid metal,

IOP Conf. Series: Materials Science and Engineering **35** (2012) 012002 doi:10.1088/1757-899X/35/1/012002

wettability and intersolubility of metals, surface energy and the temperature of liquid metal as well as the value of internal and external stresses influence the formation and the speed of development of this phenomenon. Theoretical knowledge about the mechanisms of LME formation and the influence of various parameters on these mechanisms as well as about research methods allowed the development of a test stand which made it possible to conduct tests on samples made of steels subjected to tensile stresses during hot-dip metallization.

As a result of the conducted metallization process in a zinc bath with and without an addition of tin, at various stresses applied to the samples, stress diagrams were obtained as a function of time. The analysis of diagrams for particular stress values allows the conclusion that the method of load application influences the material behaviour during metallization. Material subjected to temporary stresses during metallization is relaxed, which allows the application of greater loads. However, if we apply constant load to it, the internal stresses will accumulate and cracks in the coating and substrate will occur. The penetration of liquid metal may also take place. All these phenomena may lead to the loss of cohesive properties and consequently to a sample failure. It was also determined that at low load values the method of applying load to the prescribed stress value and the metallization time do not have a significant effect on the material behaviour after it is pulled out of the metallizing bath.

As a result of tests conducted on steel 1.0615 (C70D) in two metallizing baths, it was noticed that an addition of tin may be conducive to the occurrence of LME. During strength tests in a zinc bath, no fractures were found in the coating or in the substrate for the entire stress range examined, and the coatings obtained were continuous and uniform. However, the samples subjected to stress during metallization in a zinc bath with tin addition had fractures in the coating and in the substrate material, coupled with penetration of liquid metal. Metallographic examinations conducted with a light microscope allowed the localization of characteristic areas in the structure referred in the literature as LME.

References

[1] Liberski P, Podolski P and Mendala J 1998 *Rudy i Met. Nież.* (in polish) **11** 642
[2] Shchukin E D 1999 *Physicochemical and Eng. Asp.* **149** 529
[3] Clegg R E, Jones D R H 2003 *Eng. Failures Analysis* **10** 119
[4] Joseph B, Picat M and Barbiera F 1999 *The European Phys. J. Applied Physics* **5** 19
[5] Clegg R E 2001 *Eng. Fracture Mechanics* **68** 1777
[6] Westwood A R C, Preece C M and Kamdar M H 1971 *Academ. Press* **3** 589
[7] Stoloff N S, Davies R G and Johnston T L 1966 *Eds. Gordon and Breach*, New York 613
[8] Popovich V V, Dmukhovskaya I G 1978 *Sov. Mater. Sci.* 365
[9] Mendala J, Liberski P 2008 *Inż. Mat.* (in polish) **6** 1076

IOP Conf. Series: Materials Science and Engineering **35** (2012) 012003 doi:10.1088/1757-899X/35/1/012003

The structure of coatings obtained in the Zn-31Al-3Mg bath by the batch hot dip method

H Kania
Department of Materials Technology, Silesian University of Technology, Krasinskiego 8, 40-019 Katowice, Poland

E-mail: henryk.kania@polsl.pl

Abstract. Zn-Al dip coatings provide effective protection of steel surface against corrosion - better than traditional zinc coatings. Corrosion resistance can be further increased by adding Mg to the bath. Coatings obtained in Zn-Al-Mg baths are produced on metal plates by applying the continuous method. In this paper the author presents the results of tests on obtaining Zn-Al-Mg coatings on products with limited shape by use of the batch hot dip method. The growth kinetics of coatings obtained in the Zn-31Al-3Mg bath on steel with low silicon contents has been defined. The structure has been developed and the chemical composition of particular structural components of the coating has be established. It has been determined that the structure of coatings obtained in the Zn-31Al-3Mg bath is continuous. The course of reaction between the tested steel and liquid Zn-31Al-3Mg alloy is very abrupt, which leads to the formation of coatings with excess and non-uniform thickness.

1. Introduction

Zinc-aluminium coatings provide effective protection of steel products against corrosion. For the last few decades coatings have been produced in such galvanizing baths as: Galfan (Zn-5Al) [1] and Galvalume (Zn-55Al-1.6Si) [2]. These coatings have higher corrosion resistance than zinc coatings due to sacrificial protection provided by zinc and barrier protection provided by Al. Additional increase in the corrosion resistance of zinc-aluminium coatings is obtained by adding Mg to the bath. Corrosion resistance of coatings grows along an increase in the content of Mg within the range of 0-3wt% [3,4]. For over a dozen years the following coatings have been produced on an industrial scale: Dymazinc (Zn-0.2Al-0.5Mg) [5], Suprer Zinc (Zn-4.5Al-0.1Mg) [6], Super Dyma (Zn-11Al-3Mg-0.2Si) [3] and ZAM (Zn-6Al-3Mg) [4] and MagiZinc^{TM} (Zn, 1-2%wtMg, 1-2%Al) [7].

Coatings obtained in Zn-Al baths, including those with Mg addition, are produced only by use of the continuous method on metal plates. In this method the metal plate surface is pre-treated in controlled atmospheres. Then a hot metal plate is immersed in the galvanizing bath. The time of immersion varies within the range of a few seconds. In such conditions it is possible to obtain a continuous coating without any superficial defects. Even though Zn-Al coatings containing Mg have good properties, they are not made on products with limited shape by use of the batch hot dip method. The main restriction to use Zn-Al alloys is related with the specificity of the surface pre-preparation which includes fluxing. Adding more Al to the bath results in Al reacting with the chlorides contained in the fluxing agent which reduces its efficiency. It leads to the occurrence of coating discontinuities [8,9].

So far the technological difficulties that accompany the process of producing coatings in Zn-Al galvanizing baths have made it impossible to apply them in the industry to cover products by use of the batch hot dip method. Research on using Zn-Al-Mg alloys is not advancing either. In the paper the author has presented some possibilities of producing coatings in the Zn-31Al-3Mg bath by use of the batch hot dip method.

2. Scope and investigation methods

The scope of the carried out tests was to define the structure and growth kinetics of coatings obtained in a Zn bath with the addition of 31wt% Al and 3wt% Mg. The alloy used in the test was made of pure components and the preliminary alloy of Al-25%Mg. Thermal analysis of the alloy was carried out in order to determine the temperature of the metallizing bath. It was conducted by use of the ATD-CRYSTALDIGRAPH device. The device was equipped with a standard ceramic cup by Heraus with a NiCr-Ni thermocouple and with a transducer connected to a computer to track temperature changes in time. Liquid alloy at the temperature of 664.4°C was poured into the cup and the freezing point curve of the alloy was tracked (Fig.1). This allowed to define the temperatures of characteristic transformations which occur during the process of solidification. The recorded temperature at which solidification began was 515.4°C. In order to make sure that the range of temperatures applied in the process of metallization is safe and prevents from solidifying, the temperature of metallization had been established as 540°C.

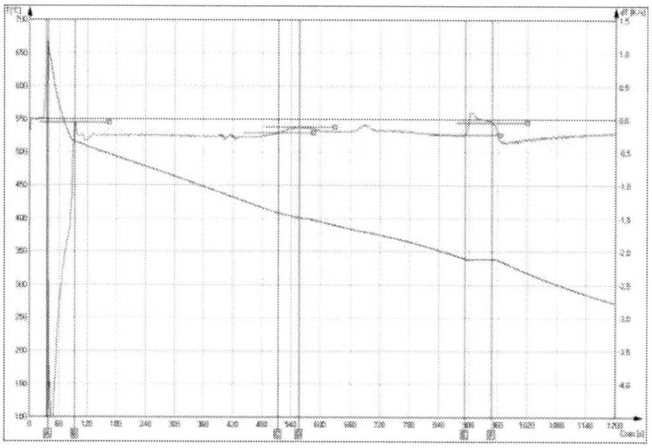

Fig.1. The graph presents changes in temperature and the first derivative for the Zn31Al3Mg alloy.

The coatings were produced on a low-silicon steel, in order to limit the influence of silicon upon the growth of the coating. The chemical composition of the steel is presented in Table 1.

Table 1. The chemical composition of steel.

Content [wt.%]					
C	Si	Mn	S	P	Fe and other
0.075	0.013	0.571	0.0159	0.0220	remainder

Steel test pieces were immersed in liquid Zn-31Al-3Mg alloy for the following times of immersion: 30, 60, 120 and 240 s. Before dipping in the bath, the surface of test pieces was prepared by degreasing, pickling and fluxing. After the pieces were removed from the bath they were cooled in the air.

In order to establish the structure of thus produced coatings, metallographic tests were performed on a light microscope. Coating thicknesses were determined by using Elcometer 355 - an electromagnetic induction-based gauge.

The chemical composition of coatings was established in the EDX analysis. The tests were performed on the Hitachi, type S-4200 scanning electron microscope with a microprobe analyser and using the Voyager 3100 software by Noram Instruments. The applied accelerating voltage was 15 kV.

3. Results

The dependence between the average total thickness of coatings obtained in the Zn-31Al-3Mg bath at the temperature of 540°C and the time of immersion is presented in Fig. 2. The obtained coatings are considerably thick. With a shorter time of immersion - 30s - the obtained coating was 149μm thick. The thickness of coating grows very abruptly as the time of immersion increases. After 240s of immersion a coating that was about 900μm thick was obtained.

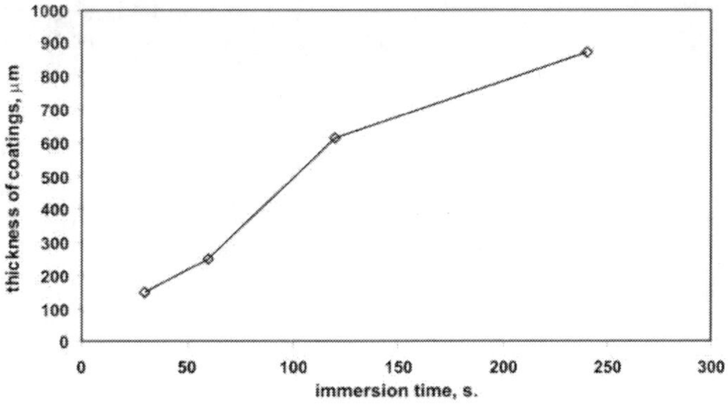

Figure 2. Growth kinetics of coatings obtained in the Zn-31Al-3Mg bath.

The cross-section of coatings obtained in the Zn-31Al-3Mg bath, with different times of immersion, is presented in Fig.3. The obtained coatings do not exhibit any discontinuity. The coating structure is non-uniform to a large degree. After a short time of immersion - 30s (Fig. 3a) - we may observe the presence of two zones: at the bases of the diffusion layer and of the outer layer. Prolonging the time of immersion to 60s results in the diffusion layer becoming stratified, which causes the occurrence of defects on the coating surface (Fig.4). As the time of immersion is further increased, the whole fragments of the diffusion layer break off, and they locate in the outer layer (Fig.3c,d). Extending the time of immersion from 60s to 240s did not cause any increase in the thickness of the compact diffusion layer at the base. It results, however, in a growth of the number of broken off particles which fosters drawing out greater amounts of alloy from the bath. In such conditions the outer layer is very thick.

The characteristic microscopic areas of the coating have been tested for their chemical composition. The image of the coating obtained on the tested steel as well as the results of the EDX analysis of selected microscopic areas of the outer layer are presented in Fig. 5. The contents of the analysed elements are presented, in percentage terms, in Table 2. The EDX analysis showed that the chemical composition of the outer layer of coating is non-uniform to a large extent. Some dark areas which are rich in Al and Zn are present in the outer layer matrix (point 1). They also contain small amounts of Mg and Fe. Those areas are surrounded by a brighter phase which contains similar amounts of Mg and Fe, but considerably less Al and more Zn (point 2). In this zone of the coating there are also some bright areas which are considerably richer in magnesium (point 3). Large particles rich in Fe, Al and Zn are visible in the coating (point 4). Their chemical composition may correspond with the Fe-Al intermetallic phase in which Zn dissolves. This phase also contains small amounts of

Mg. Particles of the Al-Zn intermetallic phase, which were the fragments of the diffusion layer that earlier broke off, come to the surface of the coating. This makes the surface uneven.

a) 30s. b) 60s. c) 120s. d) 240s.

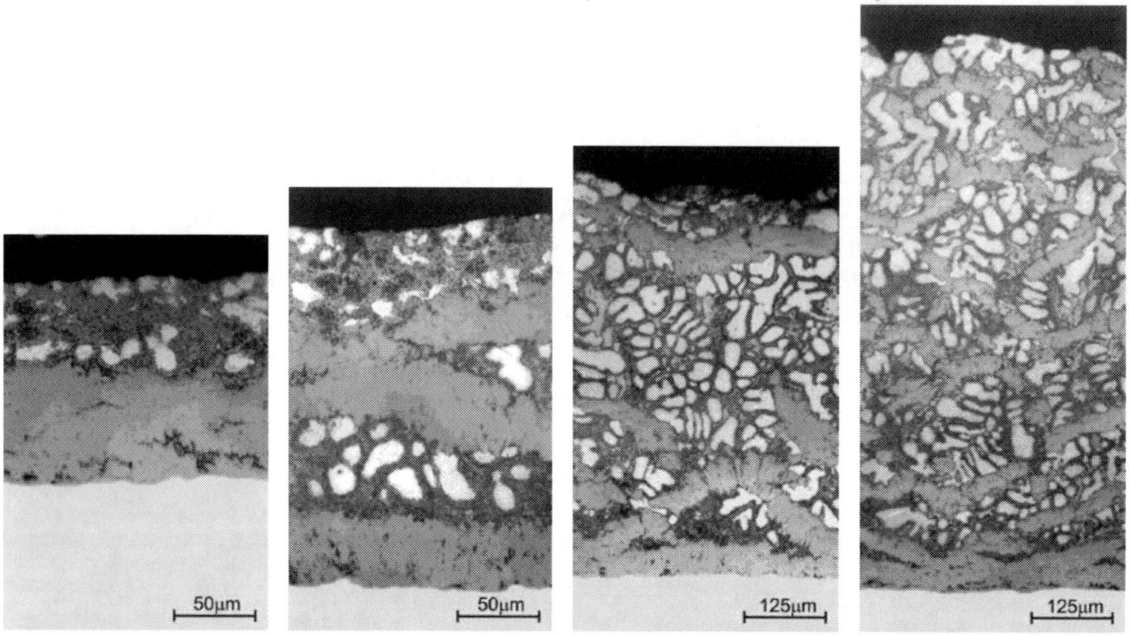

Figure 3. The structure of coatings obtained in the Zn-31Al-3Mg bath.

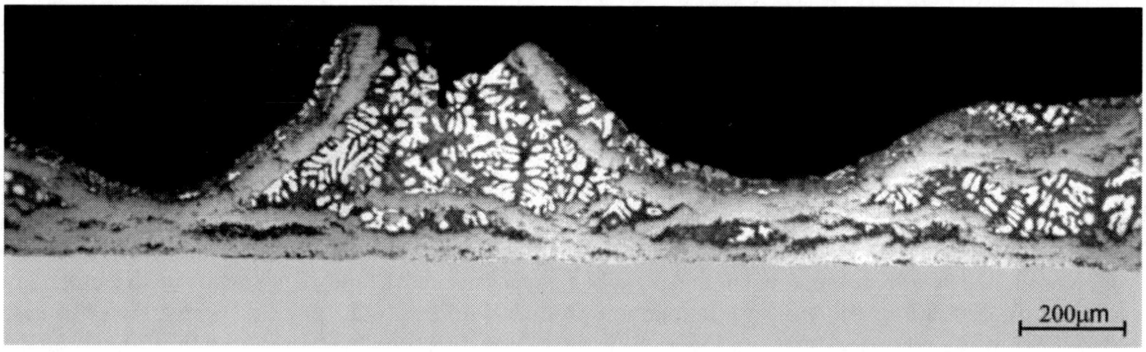

Figure 4. Defects in the structure of coating obtained in the Zn-31Al-3Mg bath, the time of immersions being 60s.

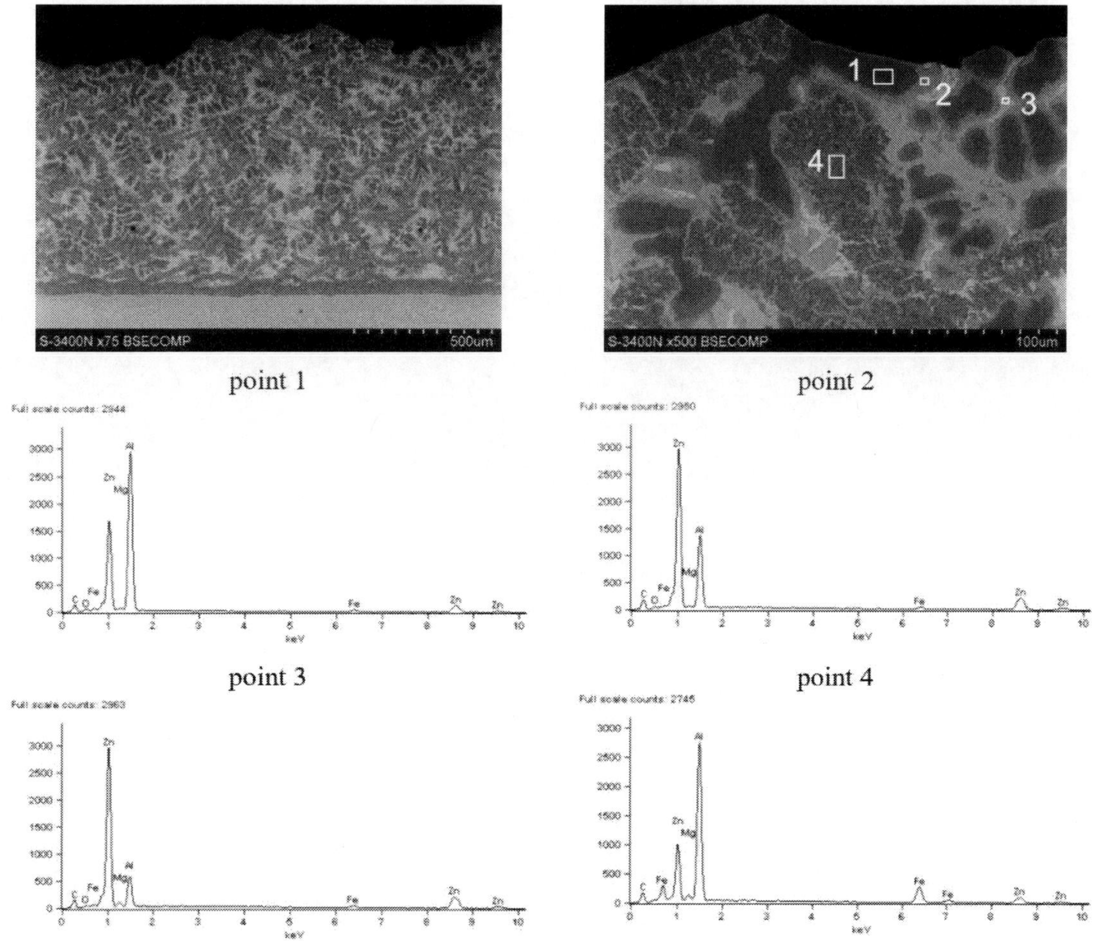

Figure 5. The microstructure of coating obtained in the Zn-31Al-3Mg bath and the EDX analysis of selected micro-areas of the outer layer of coating.

Table 2. The chemical composition of selected micro-areas of the outer layer of coating obtained in the Zn-31Al-3Mg bath (points of analysis according to Fig. 5).

Point of measurement	Element contents							
	Mg-K		Al-K		Fe-K		Zn-K	
	wt%	at%	wt%	at%	wt%	at%	wt%	at%
point 1	0.8	1.3	52.3	71.9	2.9	1.9	44.0	24.9
point 2	0.7	1.3	27.2	46.9	3.4	2.8	68.8	49.0
point 3	2.9	6.2	15.2	28.9	4.3	4.0	77.6	61.0
point 4	1.3	2.1	44.2	63.3	24.2	16.7	30.3	17.9

The microstructure of the coating's middle zone has been presented in Fig. 6. The exact structure analysis of the phase which surrounds the dark areas of the matrix indicate its non-uniform composition. Some areas with considerably higher (point 6, Table 3) and lower (point 7, Table 3) contents of Mg are present in it. The space between those structural components of the outer layer matrix are filled with a phase which contains Al and Zn and some small amounts of Fe and Mg (point 5, Table 3).

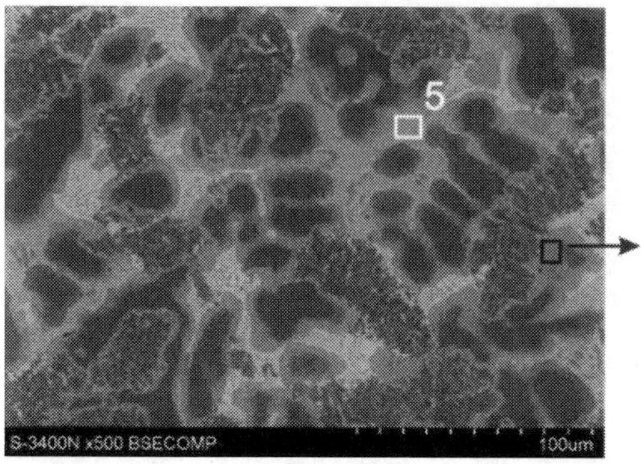

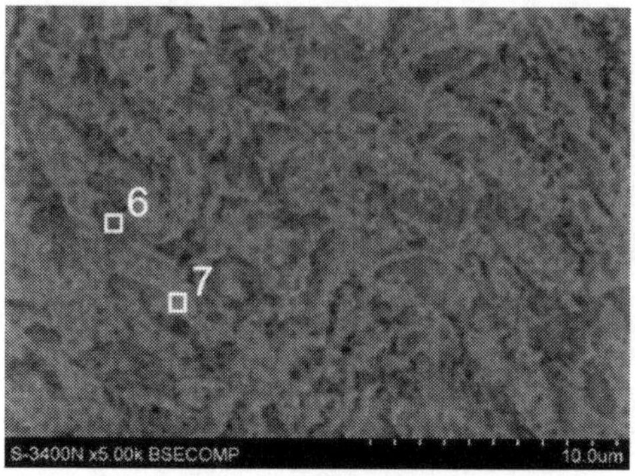

point 5

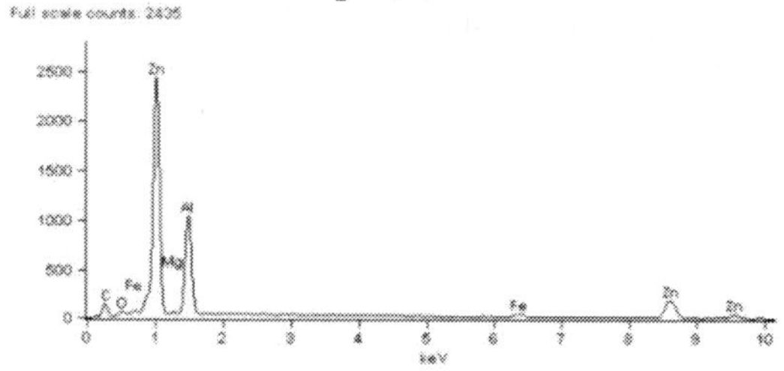

point 6

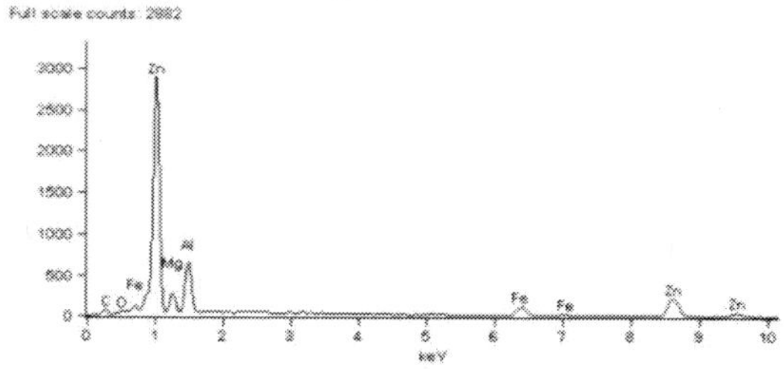

point 7

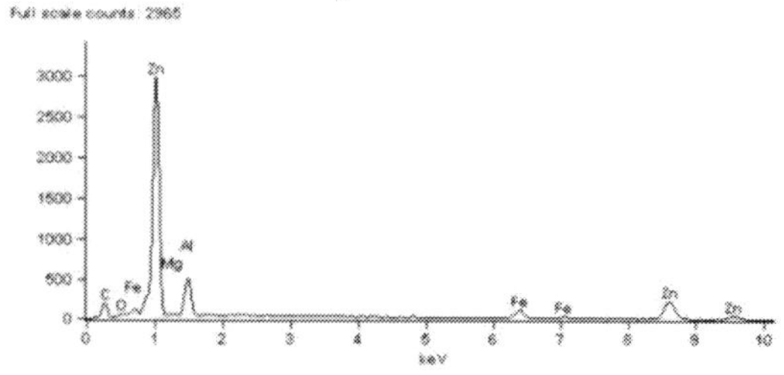

Figure 6. The microstructure of the coating's middle zone and the EDX analysis of selected microscopic areas.

Table 3. The chemical composition of selected micro-areas of the coating's middle zone (points of analysis according to Fig.6.

Point of measurement	Element contents							
	Mg-K		Al-K		Fe-K		Zn-K	
	wt%	at%	wt%	at%	wt%	at%	wt%	at%
point 5	0.8	1.6	25.3	44.4	4.3	3.6	69.6	50.3
point 6	6.7	13.4	14.8	26.8	10.0	8.7	68.5	51.1
point 7	0.9	2.1	12.8	25.5	9.4	9.0	76.9	63.4

The microstructure of coating right at the steel base is presented in Fig. 7. Based on the chemical composition analysis of some characteristic areas of the matrix (points 8, 9, 10, Table 4) it can be determined that they have a similar chemical composition to those which are present on the coating surface. The diffusion layer which is formed directly on the base contains 44.9wt% Al and 41wt%Fe and Zn (point 11, Table 4). The chemical composition indicates that this is probably the Fe-Al intermetallic phase enriched with zinc. The structure of the diffusion layer is relatively compact. It does, however, contain some small areas with quite large amounts of Mg (point 12, Table 4). This may signify that Mg solubility in the Fe-Al phase is limited.

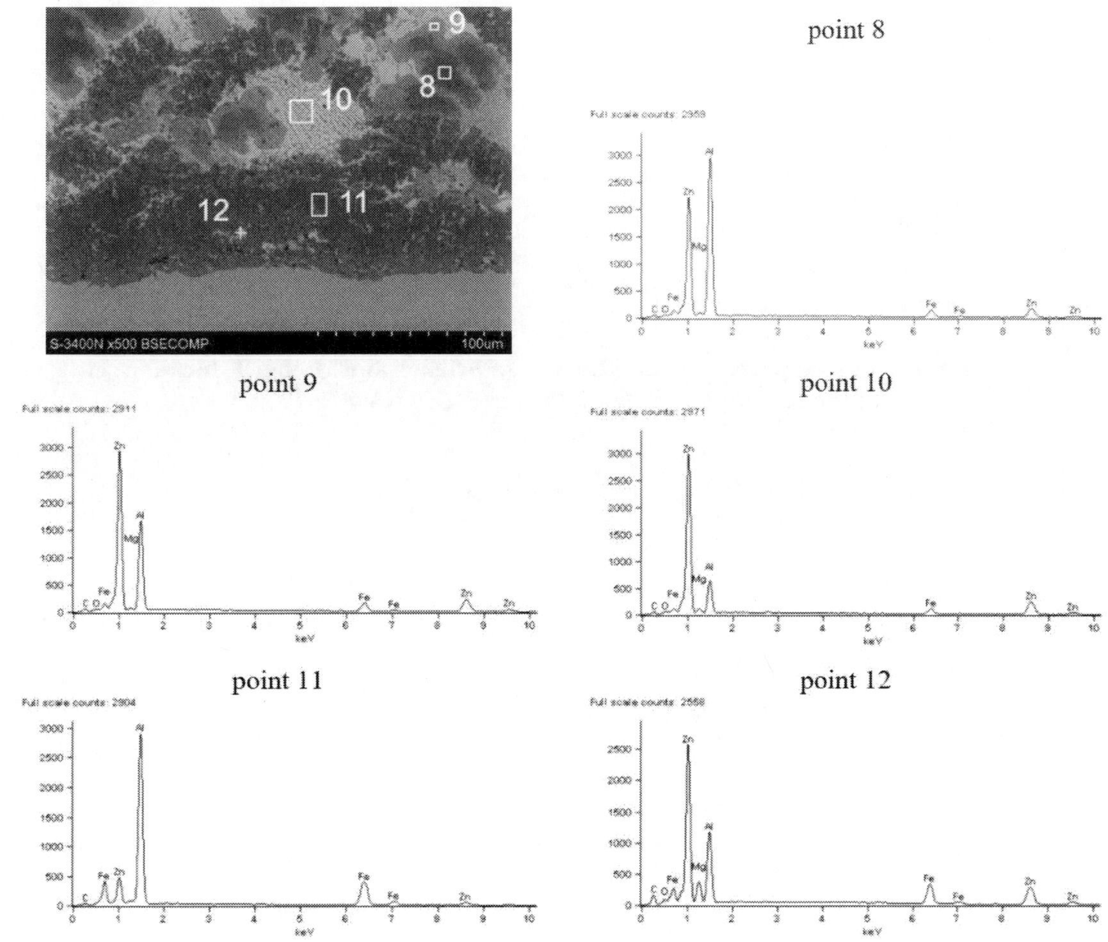

Figure 7. The microstructure of the diffusion layer of coating obtained in the Zn-31Al-3Mg bath and the EDX analysis of selected micro-areas.

Table 4. The chemical composition of selected micro-areas of the diffusion layer of coating obtained in the Zn-31Al-3Mg bath (points of analysis according to Fig. 7).

Point of measurement	Element contents							
	Mg-K		Al-K		Fe-K		Zn-K	
	wt%	at%	wt%	at%	wt%	at%	wt%	at%
point 8	0.7	1.1	45.3	65.6	10.1	7.0	43.9	26.2
point 9	0.8	1.4	28.8	48.5	11.1	9.0	59.3	41.1
point 10	2.8	6.0	14.6	27.8	9.7	8.9	72.9	57.3
point 11			44.9	63.6	41.0	28.1	14.1	8.2
point 12	6.1	11.9	16.9	29.7	19.0	16.2	58.0	42.2

4. Conclusion

Producing coatings in zinc baths with a large Al addition by using the batch hot dip method is technologically difficult. However, on the basis of the conducted tests it can be concluded that it is possible to produce a continuous coating in a zinc bath with the addition of 31wt%Al, further enriched with 3%Mg. Coatings produced at the temperature of 540°C are considerably thick and they exhibit large thickness unevenness. The differences in thickness result from the characteristics of the coating growth.

After a short time of immersion a coating composed of the diffusion layer and of the outer layer is formed. Prolonging the time of immersion causes the flaking of a part of the diffusion layer. Initially the flakes do not break off the diffusion layer but they bend towards the surface of the coating. Spaces between the flakes are filled with the metallizing bath. In such conditions some local thickness overgrowths occur. Even though the coating is continuous and it has protective properties, it is inaesthetic. Further prolonging the time of immersion causes the breaking off of fragments of the diffusion later, which fosters the formation of a very thick diffusion layer.

The conducted tests reveal great reactivity of steel in liquid Zn-31Al-3Mg alloy. The obtained coatings, though they provide protection for steel, do not have the correct appearance.

This work was supported by the Ministry of Science and Higher Education of the Republic of Poland, Project No N N508 592939

References

[1] Pelerin J, Bramaud B, Nouville J F, Coutsouradis D and Herrschaft D C 1982 *Proc.of the 13th Int. Galvanizing Conf. INTERGALVA* (London: EGGA)
[2] Borzillo A R et al., 1971 *Bethlehem Steel Corporation Japanese Patent 617971*
[3] Tanaka S, Honda K, Takahashi A, Morimoto Y, Kurosaki M, Shindo H, Nishimura K and Sugiyama M 2001 *Proc. of the 5th Int. Conf. on Zinc and Zinc Alloy Coated Steel GALVATECH* (Brussels, Belgium, pp.153-160)
[4] Tsujimura T, Komatsu A and Andoh A 2001 *Proc. of the 5th Int. Conf. on Zinc and Zinc Alloy Coated Steel GALVATECH* (Brussels, Belgium, pp.145-152)
[5] Nishimura K, Shindo H, Kato K and Morimoto Y 1998 *Proc. of the 4th Int. Conf. on Zinc and Zinc Alloy Coated Steel GALVATECH* (Chiba, Japan, pp.437-442)
[6] Tano K and Higuchi S 1985 *Nippon Steel Technical Report 25* pp.29-37
[7] Volt M, Bleeker R, Maalman T and van Perlstein E 2006 *Proc. of the Galvanized Steel Sheet Forum* (Duesseldorf: ILZRO and IZA)
[8] Porter F C 1991 *Zinc Handbook: properties processing and use in design* (Marcel Dekker, New York)
[9] Marder A R 2000 *Prog. in Mat. Sci.* **45** 191

Synergistic influence of Al, Ni, Bi and Sn addition to a zinc bath upon growth kinetics and the structure of coatings

H Kania, P Liberski
Department of Materials Technology, Silesian University of Technology, Krasinskiego 8, 40-019 Katowice, Poland

E-mail: henryk.kania@polsl.pl

Abstract. In this article the authors have analysed the current knowledge about the influence of alloy additions used in galvanizing baths. The optimum concentration of Al, Ni, Bi and Sn addition has been established. Some tests have been conducted to determine the synergistic effect of the addition of AlNiBiSn to a zinc bath upon the structure and growth kinetics of coatings. The structure of the coatings obtained on steel with low silicon contents and on Sandelin steel as well as their chemical composition have been revealed. It has been established that the addition of AlNiBiSn helps to reduce excessive growth of coating on Sandelin steel. The chemical composition and the structure of the coating on Sandelin steel are similar to the chemical composition and structure obtained on steel with regular silicon contents.

1. Introduction

The quality of zinc coatings as well as zinc consumption in hot dip galvanizing mainly depend upon steel reactivity and the intensity of molten zinc flowing off the surface of the product while it is being removed from the bath. Steel reactivity is determined by the contents of silicon and the amount of zinc that remains on a product after the bath depends on the fluidity of molten zinc. Zinc consumption is also affected by the generated waste in the form of zinc ashes and hard zinc. The correct chemical composition of the bath should allow for the possibility of affecting all the above mentioned loss-generating factors.

In order to control excessive steel reactivity, especially when the Si contents fall within the Sandelin range (0.03-0.12 wt%), nickel is added to the bath. The influence of Ni addition upon reducing the adverse effect of Si is well known [1]. The contents of Ni in the bath should not be more than 0.06wt%. Above this level Ni fosters the creation of hard zinc in the form of phase Γ_2 [2]. Such particles may be easily pulled out off the bath together with the product, thus forming defects on the coating surface. The created hard zinc in the form of phase Γ_2 is rich in Ni which causes sudden decrease in Ni concentration in the bath. If the content of Ni is even higher, it may cause an increase in the coating thickness [3]. It has also been established that lowering the level of Ni to amounts below 0.04wt% considerably diminishes its influence upon reducing steel reactivity.

An alloy addition which enhances the fluidity of molten zinc is lead, which has been used for years now. Pb has very good effect upon the process of galvanizing. However, its presence in the zinc coating has a bad influence on natural environment and on human health. That is why the EU and the USA have restricted hot dip galvanizing with Pb addition for certain applications [4,5]. Some

IOP Conf. Series: Materials Science and Engineering **35** (2012) 012004 doi:10.1088/1757-899X/35/1/012004

alternatives for lead are Bi and Sn additions. Bismuth is a relatively new addition used for zinc baths. It favours the formation of spangle without affecting the microstructure of the coating at the same time. It enhances zinc fluidity, thus improving the intensity of it flowing off the product [6]. The addition of 0.1 wt% of Bi results in a similar intensity of zinc flowing off the product surface as when we add 1 wt% of Pb [7]. Bismuth is a very stable alloy addition. It needs to be replenished proportionally to the amount of zinc added to the bath. Bismuth is not harmful to human health nor to the environment.

Bismuth addition can be introduced in the bath together with other elements. According to Fratesi [8] the Zn-0,04%Ni-0,1%Bi alloy can control the reactivity of steel with high Si contents quite effectively. Pedersen [9], on the other hand, claims that the varying contents of Ni in zinc-and-bismuth alloys do not have any considerable effect upon the microstructure and thickness of coatings.

Another element used for galvanizing baths apart from bismuth is tin. As a sole addition it does not have great influence on the quality of coating. Its high concentration favours the formation of spangle [6, 10-12]. Katiforis [13] has determined that tin, up to 3 wt%, does not have any effect upon the structure of coating. However, our own study [14] has shown that tin actually decreases the thickness of coating and slightly obstructs the growth of phase ζ already at the concentration level of 1.5 wt%. Gilles [15], on the other hand, has proven that the concentration of at least 2.5 wt% allows to reduce the effect of Si in steel. Tin is not usually used as a sole alloy addition to zinc baths.

Tests carried out with various alloy addition configurations - Zn-Ni, Zn-NiBi and Zn-NiBiSn - do not show any considerable effect of the chemical composition upon growth kinetics and the structure of coatings obtained on steel with low silicon contents [16]. They, however, allow to reduce coating thickness on reactive steel [17].

The use of bismuth and tin, and to a smaller degree of lead, as alloy additions for galvanizing baths has recently been reduced considerably. It is presumed that the presence of these elements favours the occurrence of liquid metal assisted cracking (LMAC) [18]. Liquid zinc assisted cracking concerns products with compound stress, which are mainly welded constructions. At the same time, a great majority of the manufactured zinc-coated products are those with compound and complicated shape which are not crack-sensitive. Galvanizing such products with zinc of insufficient fluidity level causes unjustified increase in zinc consumption. This is why we should not cross out the use of Bi, Sn and Pb additions to zinc baths and their choice ought to depend on the structure of the range of products to be zinc-plated.

An important alloy addition to zinc baths is Al. Aluminium improves the gloss of the coating, it reduces zinc bath surface oxidation and inhibits the formation of brittle intermetallic Fe-Zn phases and thus ensures better plasticity of coating[19]. Small amounts of Al (up to 0.005 wt%) are added in order to brighten the coating surface. Its presence causes the formation of a continuous layer of Al_2O_3 on the surface, which reduces further oxidation and functions as a barrier layer. A layer of aluminium oxide is also formed on the surface of the bath, thus protecting it against further oxidation which reduces the amount of produced ashes [20,21]. However, if its content is above 0.01 wt%, Al may react with chlorides which are components of fluxing agents, and thus reduce their efficiency.

If we analyse the present knowledge on the influence of particular elements upon the formation of zinc coatings and upon technological properties of the bath we come to a conclusion that introducing a joint addition of AlNiBiSn gives quite good results. It is one of alternative solutions used for particular applications.

2. Experimental details

The scope of the carried out tests was to define the structure and growth kinetics of coatings obtained by using zinc alloy with the addition of AlNiBiSn. As the basis for conducting the tests the authors decided to use steel containing 0.02% of silicon as low-silicon steel and steel containing 0.05% Si as steel with the critical amount of silicon from the Sandelin range. The phosphorus content in the tested steels did not exceed 0.02% which limited its impact upon the course of reaction between iron and zinc. The designation and chemical composition of the tested steels are presented in Table 1.

Table 1. The chemical composition of steel.

Designation	Content [wt%]					
	C	Si	Mn	S	P	Fe and other
0.02Si	0.138	0.021	0.743	0.0086	0.0088	remainder
0.05Si	0.065	0.051	0.87	0.0068	0.0066	remainder

The galvanizing bath was prepared of pure components. It was saturated with iron in order to limit changes in the composition of the bath during galvanizing. The chemical composition of the zinc bath is presented in Table 2.

Table 2. Chemical composition of the galvanizing bath.

Content [wt%]							
Al	Fe	Pb	Cu	Sn	Bi	Ni	Fe and other
0.0076	0.032	0.023	0.0087	0.68	0.073	0.045	remainder

Steel test pieces were galvanized at the temperature of 440°C, 450°C and 460°C and the times of immersion of 1, 3, 6 and 12 minutes. Before dipping the specimens in the zinc bath their surface was pre-prepared in a standard process. When the pieces were removed from the bath they were then cooled in the air.

In order to establish the structure of thus produced coatings, metallographic tests were performed on a light microscope. Coating thicknesses were determined by using Elcometer 355 - an electromagnetic induction-based gauge.

The chemical composition of coatings was established in the EDX analysis. The tests were performed on the Hitachi, type S-4200 scanning electron microscope with a microprobe analyser and using the Voyager 3100 software by Noram Instruments. The applied accelerating voltage was 15 kV.

3. Results and discussion

The relationship between coating thicknesses obtained on the tested steel specimens and the time of immersion is presented in Fig. 1. After analysing the influence of the time of immersion upon coating thickness it has been observed that the longer the time of immersion, the slower the increase in the coating thickness - both, for low-silicon steel (0.02Si) and for the Sandelin steel (0.05Si). An increase in liquid zinc temperature causes a growth of coating thickness for steel which falls below the Sandelin range (0.02Si). Increasing the temperature from 440°C to 460°C caused an increase of about 20μm in the coating thickness when the time of immersion was 12 minutes (Fig. 1a). Even though it is clear that the growth of temperature contributes to an increase in the coating thickness, it does not have any considerable impact upon the growth curve character. However, no impact of the zinc bath temperature upon the coating thickness for the 0.05Si steel has been observed. The coating thicknesses obtained in temperatures from 440°C to 460°C were very much similar to one another (Fig. 1b). On comparing the presented relationships it can be stated that a smaller coating thickness has been obtained for the Sandelin steel (0.05Si) than for low-silicon steel (0.02Si).

The external appearance of coatings formed in the examined Zn-AlNiBiSn bath is presented in Fig. 2. We may say that the obtained coatings do not exhibit any discontinuity. For the 0.02Si steel (Fig. 2a) the coatings are bright and shiny. The spangle is clearly visible on the coating surface. The temperature does not affect the appearance of the coating, though at the temperature of 460°C the spangle is slightly finer.

The spangle is not that visible on the Sandelin steel (0.05Si) test pieces (Fig. 2b). It can actually be observed on the surface of coating obtained at 440°C. However, at higher temperatures it starts disappearing and more and more irregularities are observed on the coating. It is worth to note that when we increase the temperature, the fluidity of liquid zinc alloy becomes greater and the outer layer of the coating becomes thinner.

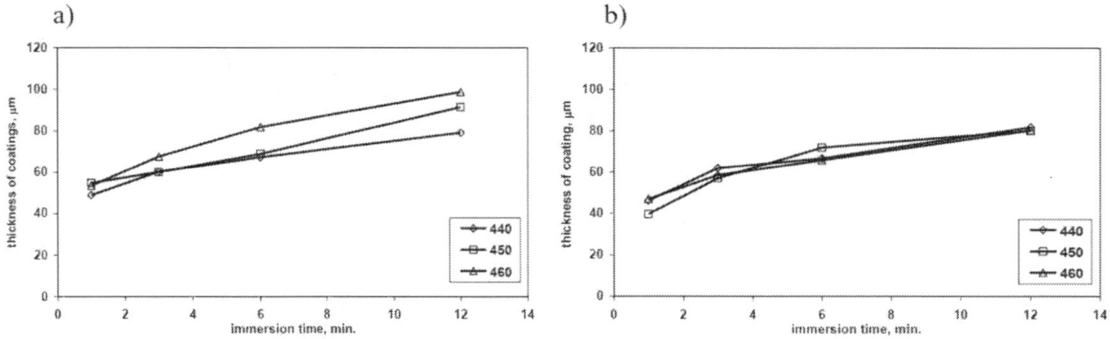

Figure 1. Growth kinetics of coatings obtained in the Zn-AlNiBiSn bath for the 0.02Si steel (a) and for the 0.05Si steel (b).

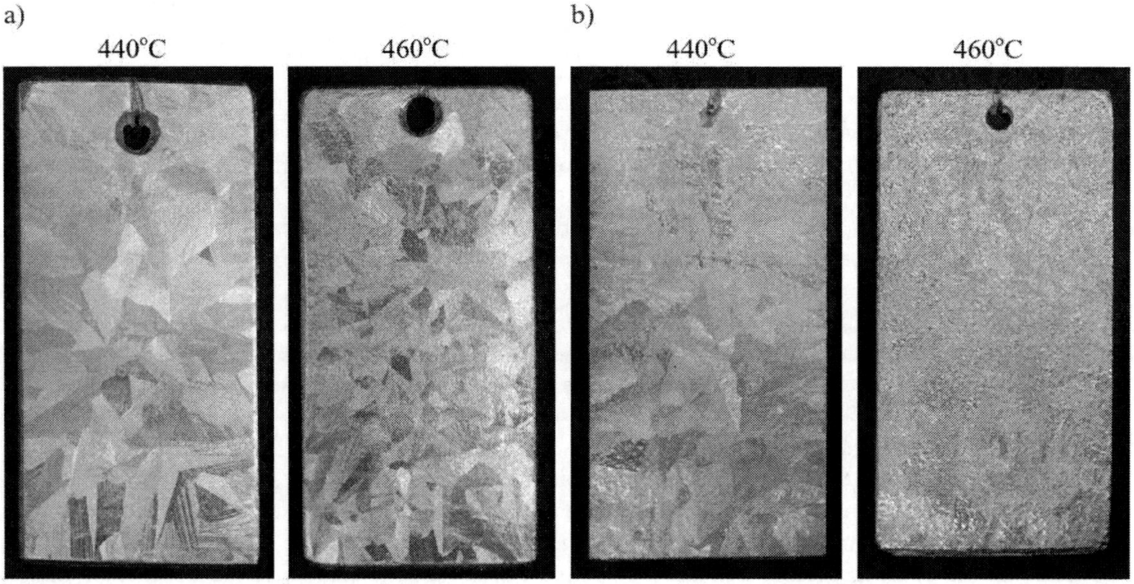

Figure 2. The appearance of the surface of coating obtained in the Zn-AlNiBiSn bath for the 0.02Si steel (a) and for the 0.05Si steel (b).

The structure of the resultant coatings has been revealed in metallographic observations performed with a light microscope. The structure of coatings obtained on the 0.02Si steel at the temperatures of 440 and 460°C and immersion times of 3 and 12 minutes is presented in Fig. 3. The appearance of the cross-section of coating obtained on steel specimens with the silicon content below the Sandelin range presents a typical, stratified arrangement of phases, starting from the steel basis, phase Γ layer - which is very thin and hard to notice if the time of immersion is short - a firm-structured layer δ_1 and layer ζ with the characteristic crystallites entering the outer layer of coating composed of Fe solution in Zn - η. The outer layer η ought to be treated here rather as a layer of the studied zinc alloy which, as a result of dissolution, always contains an amount of iron close to the saturation level. As the time of immersion is prolonged, a growth of the coating thickness can be observed. What is particularly well visible is an increase in the thickness of the firm phase δ_1 layer. It is impossible, however, to observe such a clear growth of the phase ζ layer thickness. This phase, being in direct contact with the liquid zinc alloy, is formed as a result of simultaneously occurring processes: diffuse growth and dissolving at the boundary with molten zinc. On analysing the influence of temperature on the coating structure, no significant differences have been found, although growing temperature slightly increases the thickness of the Fe-Zn intermetallic phase layers.

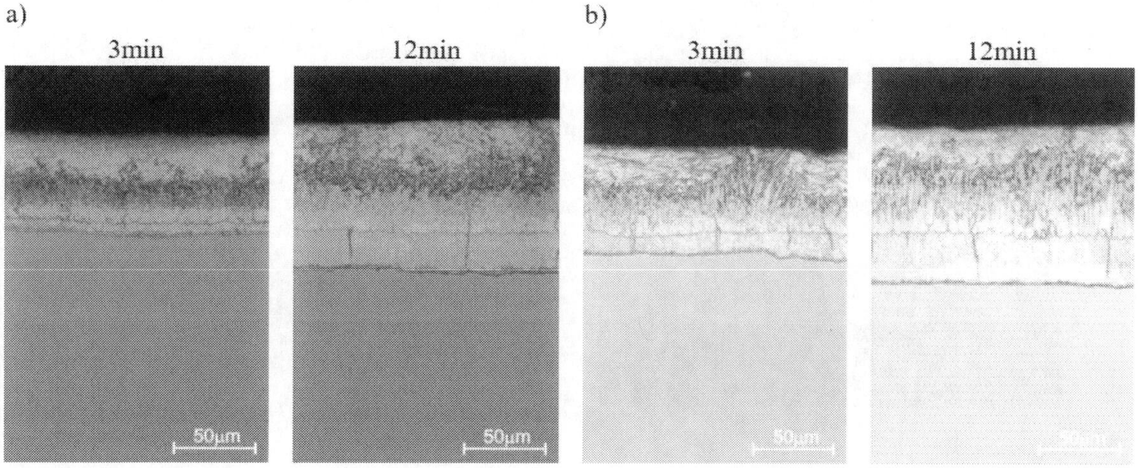

Figure 3. The structure of coatings obtained in the Zn-AlNiBiSn bath for the 0.02Si steel (a) at the temperature of 440°C (a) and 460°C (b).

The structure of coatings obtained on the 0.05Si Sandelin steel at the temperatures of 440 and 460°C is presented in Fig. 4. After analysing the structure of coating on that steel it must be stated that it is analogous to the one described for the 0.02Si hypo-Sandelin steel. We may, however, observe a smaller thickness of the outer layer - especially at the temperature increased to 460°C - which explains irregularities on the coating surface described in the coating appearance. The obtained structure is not the typical one that is usually formed on steel falling within this range of silicon concentration. When we galvanize Sandelin steel we expect to obtain a structure with highly developed diffusion layer of the coating. As a result of an intensifying effect of Si, phase ζ grows while the thickness of phase δ_1 decreases. When we analyse the structure of coating obtained in the examined Zn-AlNiBiSn bath we will find that the growth of intermetallic Fe-Zn phases in the diffusion layer is identical to that observed for hypo-Sandelin steel, which is caused by the interactions of alloying elements in the bath.

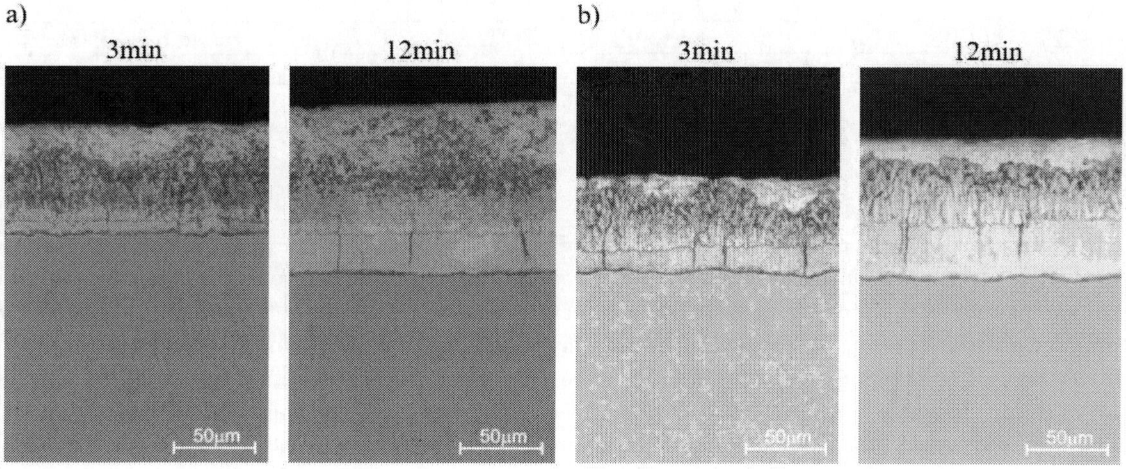

Figure 4. The structure of coatings obtained in the Zn-AlNiBiSn bath for the 0.05Si steel at the temperature of 440°C (a) and 460°C (b).

The image of coating obtained on the 0.02Si steel is presented in Fig. 5. The contents of the analysed elements in the characteristic areas are presented, in percentage terms, in Table 3. The EDX analysis showed that the outer layer of the coating mainly contains Zn (point 2). There are precipitates visible in this layer that are mainly rich in Sn and Bi (point 1). In the phase ζ layer the presence of precipitates rich in Sn between the crystallites of this phase has been established (point 3).

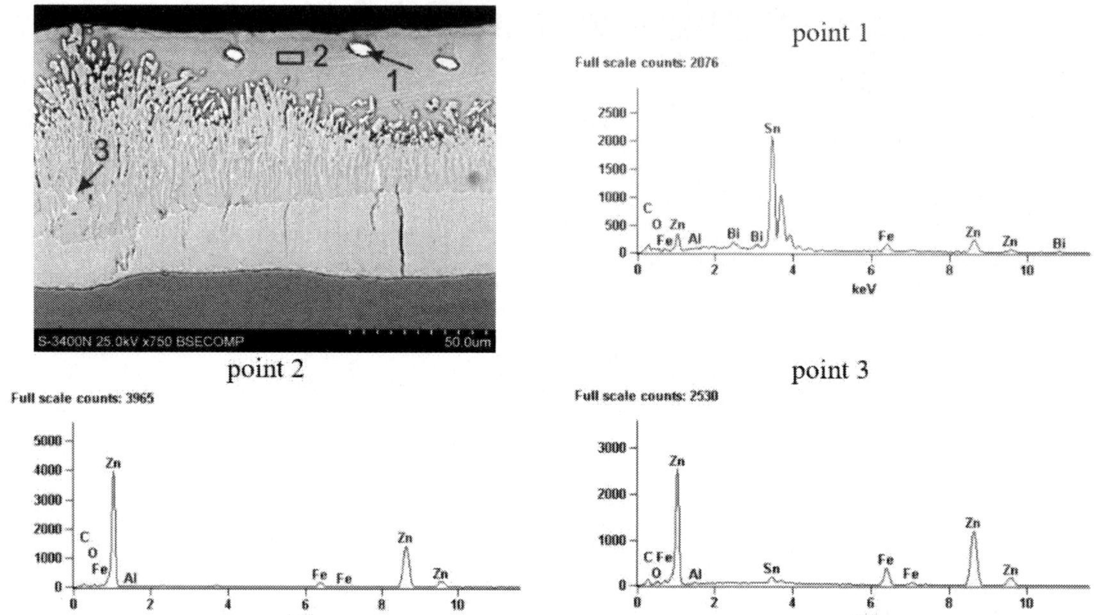

Figure 5. The microstructure of coating obtained in a Zn-AlNiBiSn bath on the 0.02Si steel and the EDX analysis of selected areas.

Table 3. The chemical composition in selected micro-areas of coating obtained in a Zn-AlNiBiSn bath on the 0.02Si steel (points of analysis according to Fig. 5 and 6).

Point of measure ment	Element contents									
	Al-K		Fe-K		Zn-K		Sn-L		Bi-M	
	wt%	at%	wt%	at%	wt%	at%	wt%	at%	wt%	at%
point 1	0.2	0.9	4.4	8.2	12.5	19.9	78.4	68.8	4.4	2.2
point 2	0.8	1.8	3.7	4.3	95.5	93.9				
point 3	0.6	1.4	10.9	12.8	81.6	82.0	7.0	3.9		
point 4	0.5	1.0	25.1	28.0	74.5	71.0				
point 5	1.0	2.4	13.5	15.5	82.4	80.5	3.1	1.7		

Fig. 6 presents the microstructure of the phase δ_1 layer. In the layer we can observe an area of compact structure δ_{1k} and an area of palisade structure δ_{1p}. In the δ_1 layer there are some cracks which propagate within area δ_{1p}. At the basis there is a layer with higher contents of Fe (point 4). We cannot, however, isolate separate areas of phase Γ and Γ_1. The interface between this layer and the phase δ_1 layer is very irregular. Layer δ_1 has a very compact structure. However, very fine precipitates can be seen in the cross-section. The analysis of the chemical composition in point 5 showed that these were tin-containing precipitates.

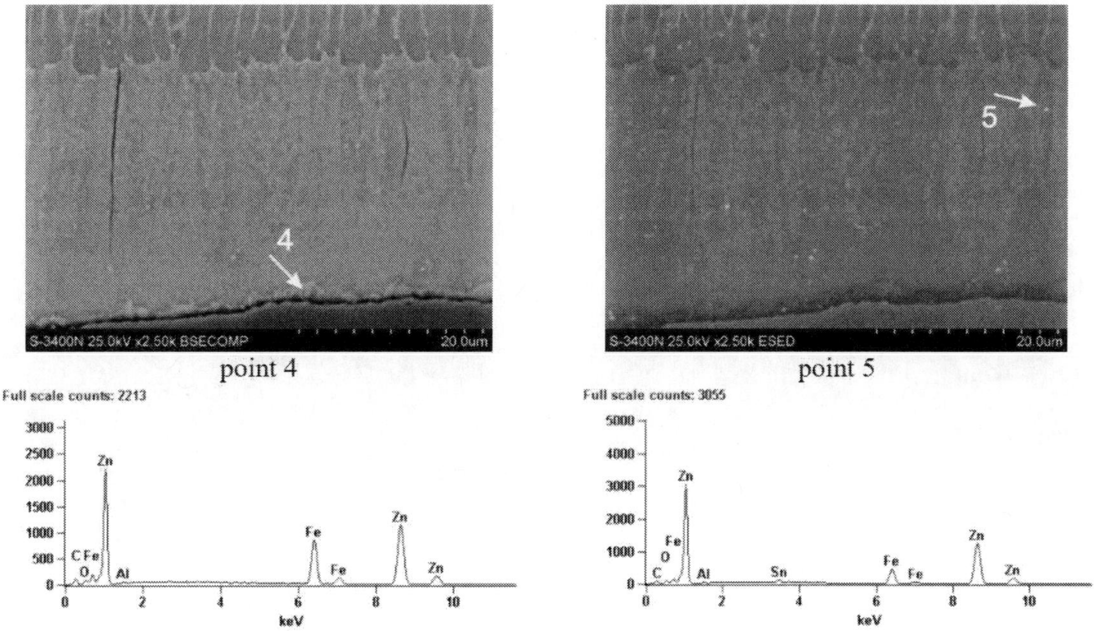

Figure 6. The microstructure of coating obtained in a Zn-AlNiBiSn bath on the 0.02Si steel and the EDX analysis of selected areas of layer δ_1.

The image of coating obtained on the 0.05Si steel is presented in Fig. 7. The contents of analysed elements in selected areas of the coating are presented, in percentage terms, in Table 4. The EDX analysis showed that the outer layer of coating mainly contains Zn (point 6). It is possible to observe Sn precipitate (point 8) and Bi precipitate (point 7) in the coating. These precipitates are not present in the outer layer but they are located at the border of phase ζ crystallites (Fig. 7). What we also note in the coating is the occurrence of an irregular and non-uniform, thin layer at the border with the basis (Fig.8). The chemical composition of particular components may point to the fact that these are phases Γ_1 at point 12 and phases Γ at point 11. The interface between phase Γ_1 and phase δ_1 is very irregular and phase Γ is not a continuous layer but it is composed of individual areas which enter the phase Γ_1 layer. Some considerable differences in Fe contents were also observed in the phase δ_{1k} layer (point 10) and phase δ_{1p} layer (point 9).

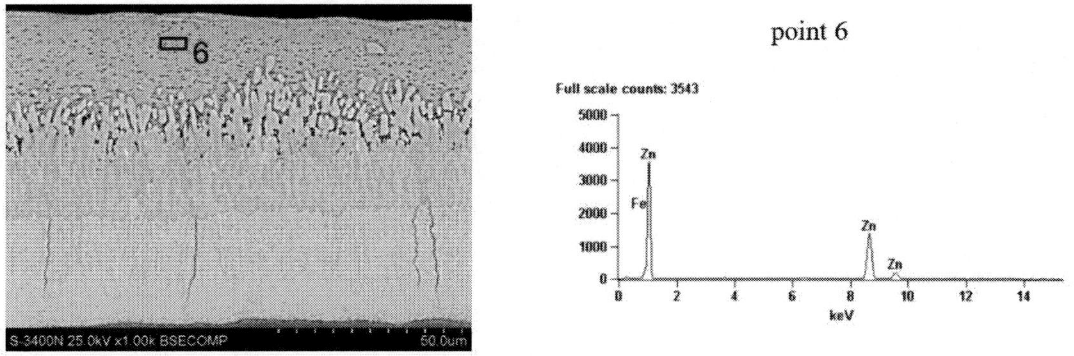

Figure 7. The microstructure of coating obtained in a Zn-AlNiBiSn bath on the 0.05Si steel and the EDX analysis of selected areas.

IOP Conf. Series: Materials Science and Engineering 35 (2012) 012004 doi:10.1088/1757-899X/35/1/012004

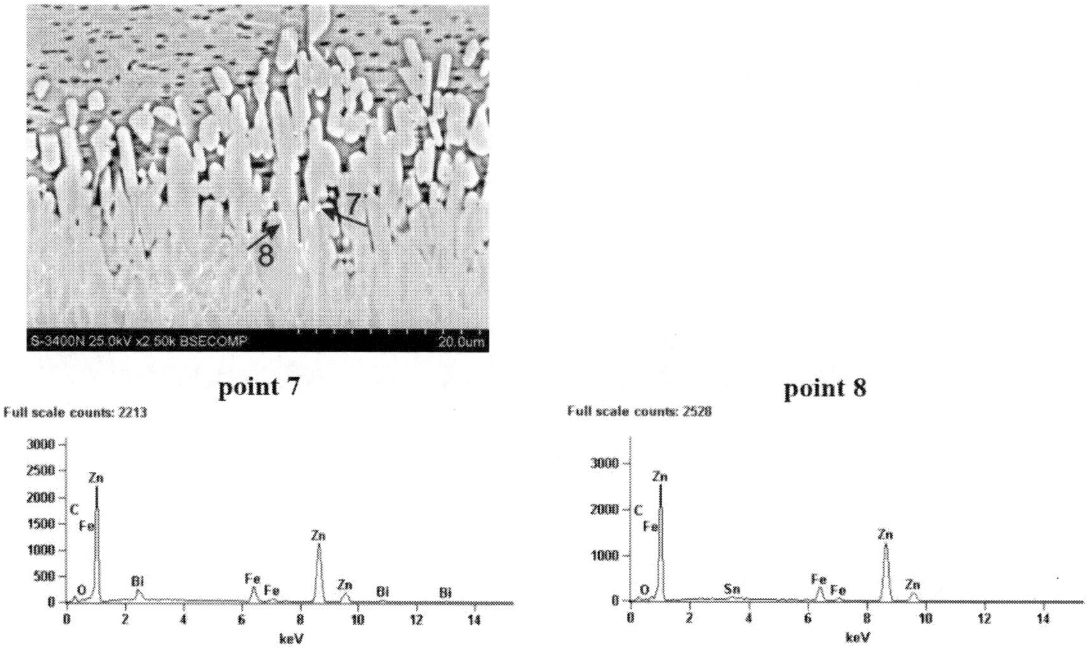

Figure 8. The microstructure of coating obtained in a Zn-AlNiBiSn bath on the 0.02Si steel and the EDX analysis of selected areas at the interface between layers η and ζ.

Table 4. The chemical composition in selected microscopic areas of coating obtained in a Zn-AlNiBiSn bath on the 0.05Si steel (points of analysis according to Fig. 6-9).

Point of measurement	Element contents									
	Al-K		Fe-K		Zn-K		Sn-L		Bi-M	
	wt%	at%	wt%	at%	wt%	at%	wt%	at%	wt%	at%
point 6			3.6	4.2	96.4	95.8				
point 7			8.7	10.9	79.2	85.0			12.1	4.1
point 8			9.4	10.9	88.2	87.7	2.4	1.3		
point 9			8.9	10.3	91.1	89.7				
point 10			11.6	13.3	88.4	86.7				
point 11	0.3	0.7	30.4	33.6	69.3	65.6				
point 12	0.1	0.3	23.9	26.8	76.0	72.8				

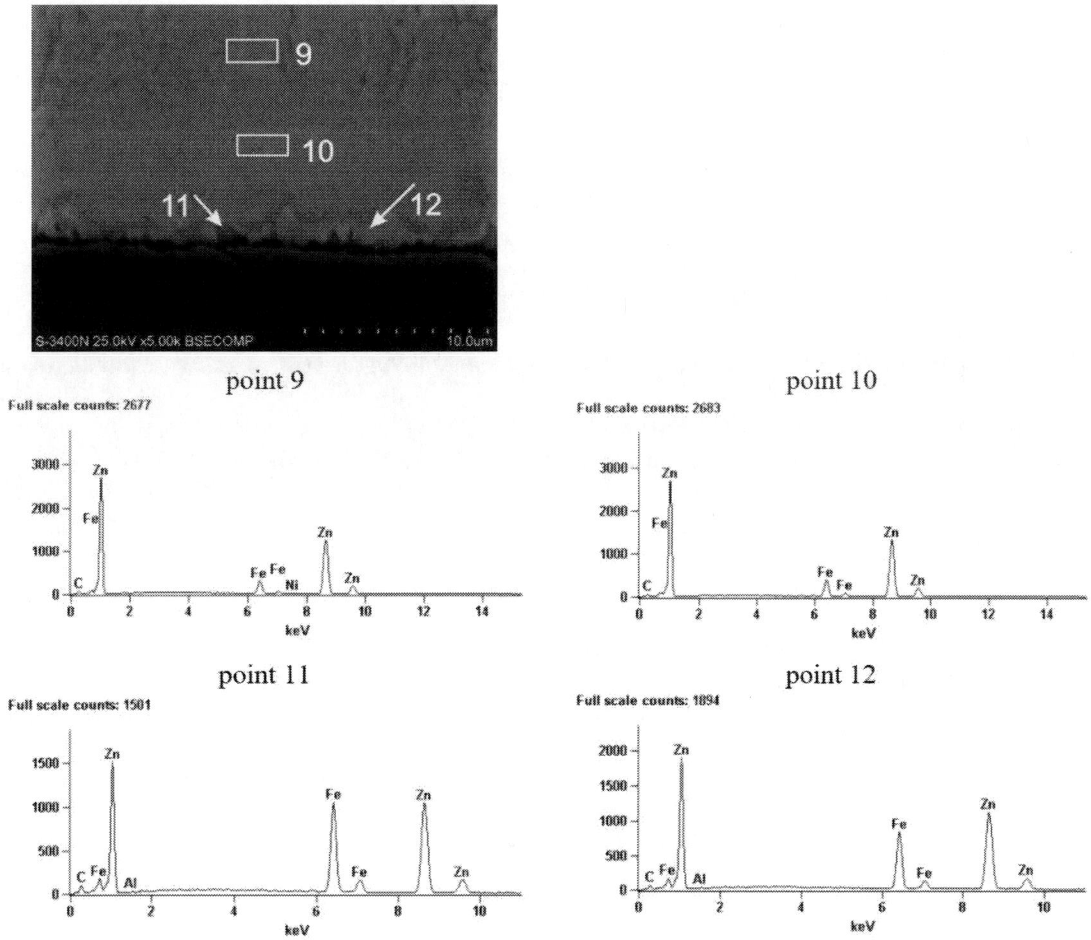

Figure 9. The microstructure of coating obtained in a Zn-AlNiBiSn bath on the 0.02Si steel and the EDX analysis of selected areas at the interface between layers δ_1 and Γ_1.

4. Conclusion

The addition of AlNiBiSn introduced jointly to a zinc bath is one of the possible chemical compositions of such a bath. This configuration of alloy additions allows to combine such profits as reduced zinc losses as a result of the bath surface oxidation, reduced formation of too thick layers, and of insufficient level of zinc fluidity.

The conducted tests provided grounds for determining the synergistic effect of the additions of Al, Ni, Bi and Sn introduced jointly to the bath upon the coating structure and growth kinetics. Coatings obtained in a Zn-AlNiBiSn bath are continuous and their appearance is regular. A spangle is formed on the surface. Its presence is caused by the effect of Bi and Sn.

Coatings obtained on the tested steel pieces were regular in structure. Alloy additions did not have any considerable effect upon coating formation on steel with hypo-Sandelin level of silicon (0.02Si). However, a reducing effect upon the Sandelin steel (0.05Si) is clearly visible. Coatings obtained on that steel are regular in structure, which is similar to the structure of coatings obtained on low-silicon steel. On analysing the coating thickness we find that coatings obtained in the examined bath on Sandelin steel are slightly thinner than coatings obtained on low-silicon steel.

The presented chemical composition of the galvanizing bath - Zn-AlNiBiSn - should be acknowledged as a composition with the optimum concentration of Al, Ni, Bi and Sn. Bismuth and tin are relatively stable additions. Aluminium and nickel become exhausted in a disproportional manner as compared to zinc consumption in the bath. This requires permanent control and replenishing the composition of the bath since it is only possible to obtain measurable effects if we maintain the concentration of particular components within optimal limits.

References

[1] Reumont G, Perrot P and Foct J 1998 *J. Mat. Sci.* **33**, 4759
[2] Tang N Y *J.Phase Equilibria* 1995 **16(2)** 110.
[3] Kania H, Liberski P and Podolski P 2005 *Material Engineering* (in polish) **5(147)** 763
[4] Beguin P, Bosschaerts M, Dhaussy D, Pankert R and Gilles M 2000 *Proc. of 19th Int. Galvanizing Conf. INTERGALVA* (Berlin: EGGA)
[5] Reumont G and Perrot P 1994 *Proc. of 17th Int. Galvanizing Conf. INTERGALVA* (Paris: EGGA)
[6] Radeker W and Friehe W 1964 *Proc. of 7th Int. Conf. on Hot Dip Galvanizing* (Paris: Pergamon Press) pp.167-178.
[7] Gagne M 1997 *Proc. AGA Tech Forum* (Nashville: AGA)
[8] Fratesi R, Ruffini N, Malavolta M and Bellezze T 2002 *Surf. Coat. Technol.* **157** 34
[9] Pedersen J, Lewis G P and Pedersen J G 1999 *Proc. of 4th Asian Pacific General Galvanizing Conf.* (Kuala Lumpur)
[10] Fasoyino F A and Weinberg F 1990 *Met. Trans. B-Process Met.* **21**, 549
[11] Sebisty J J and Palmer R H 1961 *Proc. of 6th Int. Conf. on Hot Dip Galvanizing* (Interlaken: ZDA)
[12] Chang S and Shin J C 1994 *Corros. Sci.* **36** 1425
[13] Katiforis N and Papadimitriou G 1996 *Surf. Coat. Technol.* **78** 185
[14] Kania H and Liberski P 2008 *Material Engineering* (in polish) **3**, 149
[15] Gilles M 1997 *Proc. of 18th Int. Galvanizing Conf. INTERGALVA* (Birmingham: EGGA)
[16] Kania H, Liberski P, Podolski P and Tatarek A 2007 *Ores and Non-Ferrous Metals* (in polish) **6** 348
[17] Kania H, Liberski P and Podolski P 2006 Metallurgy-Metallurgical Engineering News **5(73)** 233
[18] Huckshold M 2009 *Proc. 22th Int. Galvanizing Conf.* (Madrid: EGGA)
[19] Marder A 1990 *Microstructural characterization of zinc coatings* (TMS Warrendale, PA) 55
[20] Porter F C 1991 *Zinc Handbook: properties processing and use in design* (ed. Marcel Dekker, New York)
[21] Marder A R 2000 *Prog. Mat. Sci.* **45** 191

IOP Conf. Series: Materials Science and Engineering **35** (2012) 012005 doi:10.1088/1757-899X/35/1/012005

Phase transformations in CMSX-4 nickel-base superalloy

A Szczotok, R Przeliorz
Silesian Technical University, Faculty of Materials Engineering and Metallurgy, Krasinskiego 8, 40-019 Katowice, Poland

E-mail: agnieszka.szczotok@polsl.pl

Abstract. Thermophysical properties (including transformation temperatures) are the critical input parameters in mathematical models of solidification and casting of metallic materials. The technique of Differential Scanning Calorimetry (DSC) was selected for the study of precipitation and phase changes in CMSX-4 alloy system. The results of the calorimetric measurements together with phase transformation temperatures determined from DSC heating and cooling curves was described. In the present paper, DSC results accompanied by scanning electron microscope characterization of CMSX-4 superalloy is applied to understand significance of precipitation or phase changes in alloy system for design heat treatment for the nickel based superalloy

1. Introduction

Nickel-based superalloys are well known for their many high temperature applications, including their use in critical gas turbine components in aerospace and in land based power generation. At high temperatures, superalloys are subjected to severe operating environments that include creep, thermo mechanical stresses, oxidation and corrosion. This could result in the development of creep damage, thermal fatigue cracks and surface degradation but superalloys offer excellent high temperature tensile strength, stress rupture and creep properties, fatigue strength, oxidation and corrosion resistance as well as a microstructural stability at elevated temperatures [1-6]. A variety of strengthening mechanisms, such as precipitation hardening and solid solution strengthening, are employed for the manufacture of CMSX-4 superalloy in order to obtain its desirable microstructure which will ensure appropriate properties. The studied superalloy, developed using a multi-dimensional approach over a 10-year period, achieves a high level of balanced properties. It owes its excellent high temperature strength to the large amount of a finely dispersed ordered phase γ' [7].

Thermophysical properties (including transformation temperatures) are the critical input parameters in mathematical models of solidification and casting of superalloys. For highly complex materials like nickel-base superalloys a complete thermodynamic description of the system involves a huge amount of work, and the predicted values may differ from those measured, because existing property data are sometimes only available for selected pure elements.

The technique of Differential Scanning Calorimetry (DSC) has been applied for nickel-based superalloys investigations [8-12]. Unfortunately, indispensable data for CMSX-4 heat treatment planning is ambiguous, incomplete and unrepeatable [9,13]. In the present paper well-established DSC technique was selected for the study of precipitation and phase changes in CMSX-4 nickel-based superalloy system. Application of the obtained results will enable to design CMSX-4 superalloy heat treatment and the material after successive stage of the heat treatment will be used for a precise quantitative analysis of γ' phase precipitates occurring in it. The quantitative evaluation of the precipitates is difficult as a result of their complex morphology and their occurring in the γ+γ' eutectic. The total area fraction of γ' phase a relevant measuring scheme and formula have been suggested [14-15].

2. Material

A second-generation (approximately 3 wt. % of rhenium), single-crystal CMSX-4 nickel-base superalloy is used in the present work. The chemical composition of the superalloy is given in Table 1. CMSX-4 contains a high concentration of refractory elements, only minimal carbon.

Table 1. Chemical composition of the CMSX-4 nickel-base superalloy, in wt. %.

Ni	Cr	Re	Co	Mo	W	Ta	Ti	Al	Si	Mn	Others
balance	6.4	2.9	9.6	0.61	6.4	6.5	1.02	5.61	0.4	0.1	0.05 Zr, 0.1 Hf, 0.042 Fe, 0.0037 C

To visualize the γ-γ' microstructure 10g $CuSO_4$+50 ml HCl+50 ml H_2O etchant is applied for metallographic preparation. The etchant dissolves the γ' phase. The solidified microstructure of the investigated superalloy is shown in Figure 1. As expected, the superalloy exhibits highly segregated dendritic arms. A relatively fine precipitates of γ' phase are distributed within the dendrite cores and they become coarse near the cellular eutectic γ-γ' at the interdendritic region are visible in the microstructure. The presented images are registered using Olympus GX71 light microscope (Figures 1a, 1b) and FE SEM HITACHI 4200 equipped with X-ray spectrometer VOYAGER 3500 of NORAN (Figures 1c, 1d).

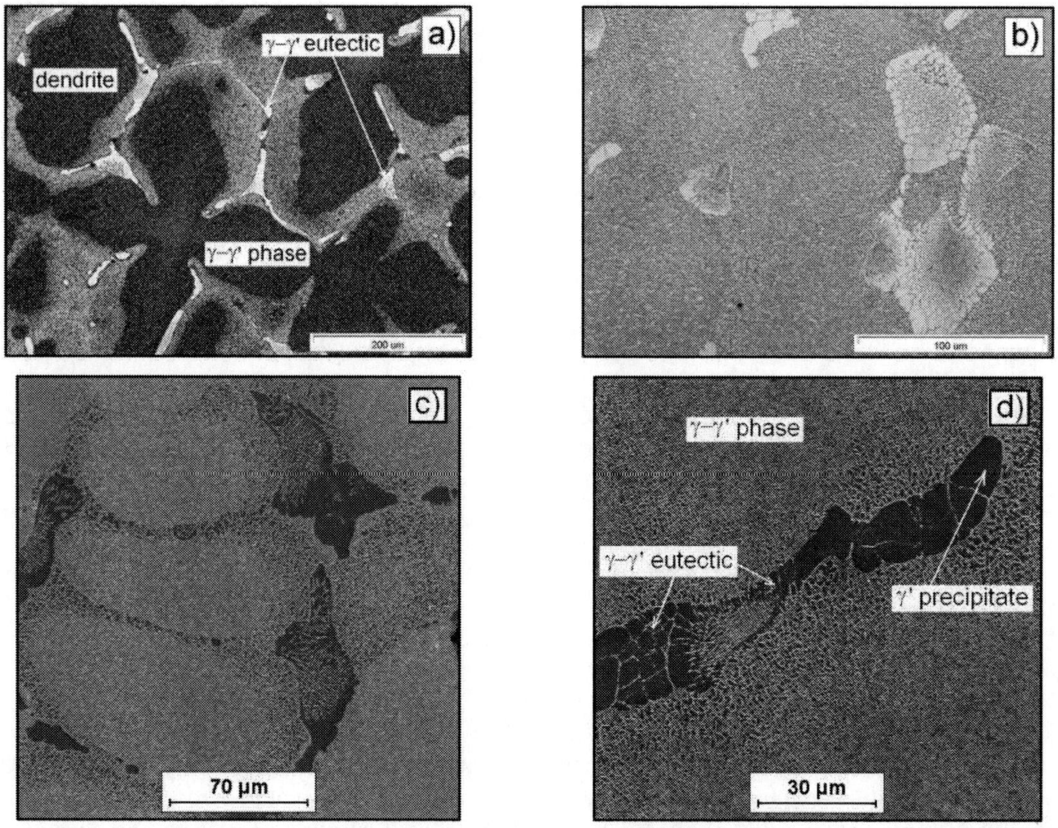

Figure 1. Micrographs of as-cast CMSX-4 superalloy: a)÷b) LM images and c)÷d) SEM images. Sample etched with 10 g $CuSO_4$+50 ml HCl+50 ml H_2O solution.

IOP Conf. Series: Materials Science and Engineering **35** (2012) 012005 doi:10.1088/1757-899X/35/1/012005

The microstructure of CMSX-4 superalloy can be optimized to be fully solutionized, to contain neither γ+γ'eutectic phase nor regions of incipient melting, carbides or microporosity. Traditionally, two heat treatments are applied for CMSX-4 nickel-based superalloy. First is the solution heat treatment, designed to homogenize the microstructure and reduce the effects of elemental segregation. The second is aging heat treatments, designed to develop a cuboidal γ/γ' microstructure.
To propose the homogenization-solution heat treatment as well as aging for the studied CMSX-4 superalloy we need to determine the liquidus, solidus and γ' solvus for the phase system.
The heat-treatment window allows for improved microstructural control since complete dissolution and re-precipitation of the fine γ' could be achieved during heat treatment.

3. Experimental

High temperature transformations of as-cast CMSX-4 SC superalloy were investigated by SETARAM Multi HTC differential scanning calorimetry (DSC) instrument. The heat/cooling rate was 5°C/min. The experiments were carried out within temperature ranging from 25 to 1450 °C. After reaching the temperature of 1450 °C (after melting) the samples were control cooled to 25 °C. The all tests were performed in the protective atmosphere of inert argon. The argon gas atmosphere applied was of high-purity 6.0 grade (≥ 99.9999%). The Al_2O_3 powder was a reference material.
The DSC specimens with a mass 85mg were cut from the CMSX-4 SC bar to the required size: cubic with 3x3 mm side. After degreasing the sample of the CMSX-4 alloy was analyzed in Al_2O_3 crucible with volume of 0.45 cm^3 partly filled with Al_2O_3.
A calibration of the DSC instrument has been performed on the basis of the melting temperature of pure Ni.

4. Results and discussion

The DSC curves obtained during heating and cooling were presented adequately in Figures 2 and 3.

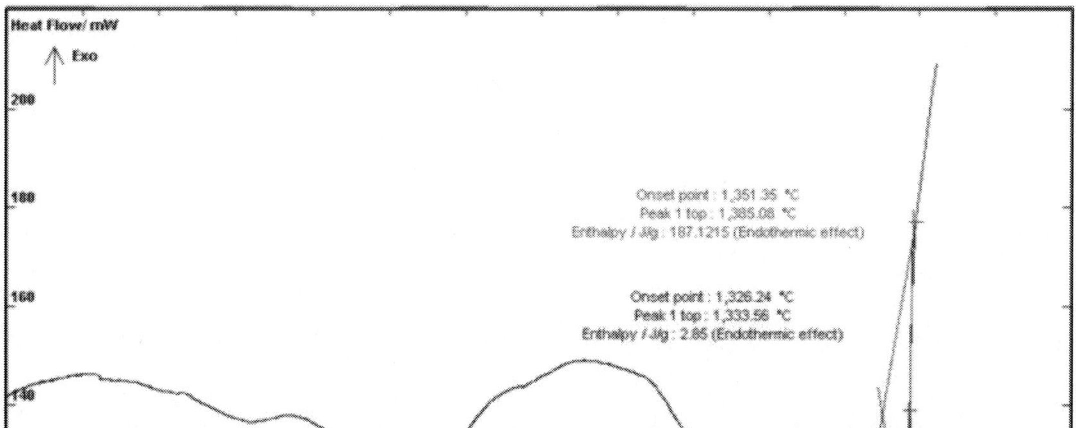

Figure 2. The DSC curve of CMSX-4 superalloy obtained during heating at 5°C/min with marking of characteristic temperatures of phase transformations.

The application of slow heating and cooling rates enabled to minimize the temperature errors, but a compromise must be arrived at because if rates are too slow the peaks produced may not be observable. The heating/cooling rates as well as the sample size during the carried out measurements were selected on the basis of own experiences of authors and examples described in the literature.

In order to monitor the phase transformation temperatures during melting and solidification, the data obtained on heating and cooling were used to determine the liquidus and solidus temperatures (Table 2). The liquidus temperature was determined at the minimum of endothermic effect at heating. During cooling the liquidus temperature was extrapolated temperature of peak start (T onset). The method of determination of temperatures of phase transformations was described in the works [16-19].

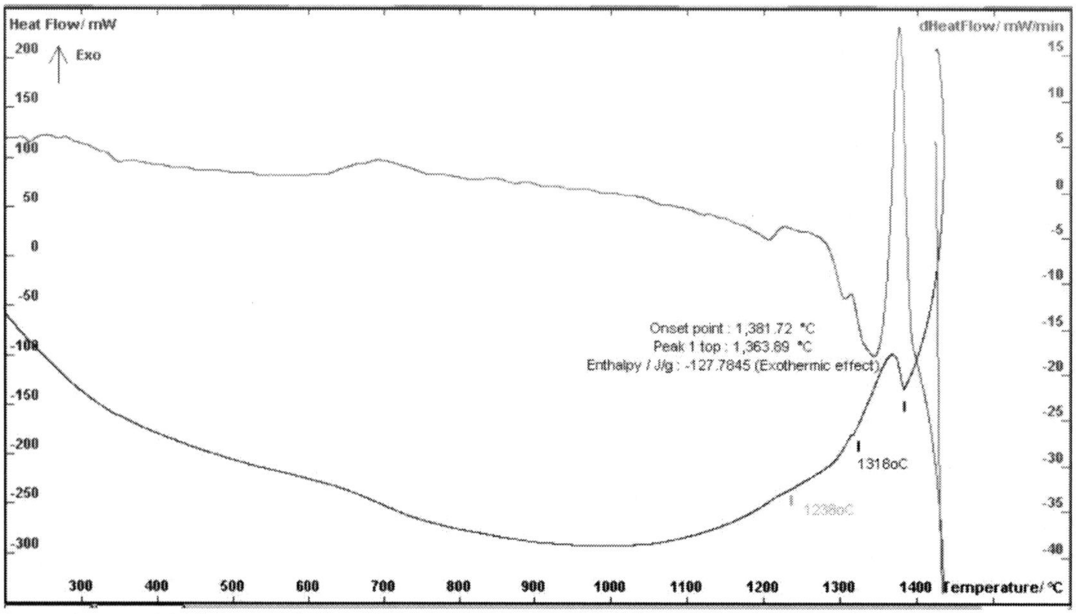

Figure 3. The DSC curve of CMSX-4 superalloy obtained during cooling at 5°C/min with marking of characteristic temperatures of phase transformations and the curve of dHF/dt derivative.

Table 2. Transformation temperatures of the CMSX-4 superalloy as the results of DSC investigation.

	Temperature of phase transformation (°C)			
	liquidus	**solidus**	**γ' solvus**	**γ' precipitation**
Heating rate 5°C/min	1385	1326	1277	
Cooling rate 5°C/min	1381	1318		1238

The temperature correction was found to be -2°C on the basis of the melting temperature of pure Ni.

5. Conclusion

The project of homogenization–solution heat treatment for nickel-based superalloy with many refractory elements, like CMSX-4, is not easy because of the following reasons: (1) a highly segregated solidified microstructure may have a local solidus temperature in the interdendritic region significantly lower than the average solvus temperature within the dendrites; (2) identification of the solvus and solidus temperatures of as-solidified superalloy from the differential scanning calorimetry (DSC) profiles is difficult because of broadening of the corresponding endothermic and exothermic peaks caused by chemical heterogeneity; (3) the complete dissolution of eutectic γ+γ' is difficult because of γ' coarse blocky morphology as well as a low surface area of the coarse γ' precipitates per volume ratio which makes it dissolve very slowly into the γ matrix.

The liquidus, solidus and γ' solvus were determined and reported. The temperature of the γ' precipitation was determined on the DSC cooling data since the solvus temperature of the γ' phase could be determined more accurately than based on heating data. During the control heating of the CMSX-4 superalloy three major phase transformations were revealed. The following temperatures of phase transformations were determined (Figure 2, Table 2): T_L (liquidus) = 1385°C, T_S (solidus) = 1326°C and $T_{\gamma' SOL}$ (γ' solvus) = 1277°C. The analyzed CMSX-4 superalloy is characterized by a high gamma prime solvus temperature.

During the control cooling of the investigated superalloy also three major phase transformations were revealed. The following temperatures of phase transformations were determined (Figure 3, Table 2): T_L (liquidus) = 1381°C, T_S (solidus) = 1318°C and $T_{\gamma' PRE}$ (γ' precipitation) = 1238°C.
The determined values of the temperatures of the phase transformations are similar to those described in the work [20].

A measurement of smaller transaction such as a eutectic reaction can be detected by the DSC method only when the cumulative heat effect exceeds the detection limit of the instrument. The temperature of the transformation can be affected by undercooling. During the carried out measurement the γ' eutectic transformation temperature was not observed.

However, it should be considered that chemical composition of CMSX-4 superalloy according to the standard concerns contents range of the elements and in the range the acceptable variation of the elements changes of the transformation phase temperatures can be present.

Acknowledgements
This work was financially supported by Polish Ministry of Science and Higher Education – Grant Iuventus Plus No 0407/T02/2010/70.

References

[1] Bradley E F 1969 *Source book on materials for elevated – temperature applications* (American Society for Metals, Metals Park, Ohio, USA).
[2] Sabol G P and Stickler R 1969 *Physica Status Solidi* **35(1)** 11.
[3] Sims C T, Stoloff N S and Hagel WC 1987 *Superalloys II High-temperature Materials for Aerospace and Industrial Power* (John Wiley & Sons Inc, New York, USA).
[4] Jena A K and Chaturvedi M C 1984 *J. Mater. Sci.* **19** 3121.
[5] Bradley E F 1988 *Superalloys: a technical guide* (Metals Park, OH, ASM International, USA)
[6] Kotval P S 1969 *Metallography* **1 (3-4)** 251.
[7] Harris K, Erickson G L, Schwer R E, Frasier D J and Whetstone J R 1990 Process and alloy optimization for CMSX-4 superalloy single crystal airfoils, in: Bachelet E. et al. (Eds.) *High Temperature Materials for Power Engineering Part II*, Proceedings Cost Conf. Liege, 24-26 September Kluwer Academic Publishers, Dordrecht, Holland, 1281.
[8] Sidhu R K, Ojo O A and Chaturvedi M C 2008 *J. Mater. Sci.* **43** 3612.
[9] Chapman L A 2004 *J. Mater. Sci.* **39** 7229.
[10] El-Bagoury N and Nofal A 2010 *Mater. Sci. Eng.* **A 527** 7793.
[11] Binczyk F, Śleziona J, Cwajna J and Roskosz S 2008 *Arch. Foundry Eng.* **8** No 3 5.
[12] Han S, Zhao K, Langhong L 2009 Solidification behavior of a nickel-base Re-containing superalloy for conventional casting Chinese Journal of Materials Research vol 23 No 1 pp 98-102
[13] Walter C, Hallstedt B and Warnken N 2005 *Mater. Sci. Eng.* **A 397** 385.
[14] Szczotok A, Richter J and Cwajna J 2009 *Mater. Charact.* ***60*** 1114.
[15] Szczotok A, Rzychoń T, Cwajna J 2010 *Inżynieria Materiałowa.*
[16] Wu R I and Perepezko J H 2000 *Metallurg. Mater. Trans.* **31A** 497.
[17] Zupanic F, Boncina T, Krizman A and Tichelaar F D 2001 *J. Alloys Comp.* **329** 290.
[18] Dočekalová S, Dobrovská J, Smetana B 2008 The effect of cooling rate on the phase transformation temperatures of IN7925A, *Metal*, 13-15 May, Hradec/Moravici, Czech Republic, 1.
[19] Fuchs G E 2001 *Mater. Sci. Eng.* **A300** 52.
[20] Wilson B C, Hickman JA and Fuchs G E 2003 *JOM* ***3*** 35.

IOP Conf. Series: Materials Science and Engineering **35** (2012) 012006 doi:10.1088/1757-899X/35/1/012006

Microstructural studies of carbides in MAR-M247 nickel-based superalloy

A Szczotok, K Rodak
Silesian Technical University, Faculty of Materials Engineering and Metallurgy, Krasinskiego 8, 40-019 Katowice, Poland

E-mail: agnieszka.szczotok@polsl.pl

Abstract. Carbides play an important role in the strengthening of microstructures of nickel-based superalloys. Grain boundary carbides prevent or retard grain-boundary sliding and make the grain boundary stronger. Carbides can also tie up certain elements that would otherwise promote phase instability during service. Various types of carbides are possible in the microstructure of nickel-based superalloys, depending on the superalloy composition and processing. In this paper, scanning electron and scanning transmission electron microscopy studies of carbides occurring in the microstructure of polycrystalline MAR-M247 nickel-based superalloy were carried out. In the present work, MC and $M_{23}C_6$ carbides in the MAR-M247 microstructure were examined.

1. Introduction

Addition of carbon in amounts from 0.05 to 0.2% and carbide forming elements enables precipitation of mainly $M_{23}C_6$, MC and M_6C types carbides in the superalloys. These carbide phases are not stable. Under the influence of service temperature and time they may undergo transformations changing their kind, size and morphology, which affect the alloys' properties at high temperatures [1].

The carbide phase precipitates have an important role on the mechanical properties of superalloys [2-5]. Discrete carbides precipitating at grain boundaries (GBs) can inhibit GB sliding and improve the strength and creep resistance of the superalloy at elevated temperatures. Primary, as well as secondary carbides, strengthen superalloys making difficult grain boundary slip during creep and tying up elements that could contribute to phase instability during exploitation of the alloy into stable phases [6]. Grain boundary slip is hindered by $M_{23}C_6$ carbides in favor of creep resistance. Morphology of these carbides has a significant effect on nickel-based superalloys properties. Shape change of $M_{23}C_6$ carbides along grain boundaries from blocky to fine, globular particles leads to increase of temporary creep resistance and improvement of nickel-based superalloys plasticity [1].
The formation sequence, composition and morphological evolution of MC carbide, from slow to rapid solidification conditions, have been characterized in numerous works [7-10].

In general, the morphology of MC carbides in an as-cast Ni-based superalloy mainly depends on its composition and casting parameters because this type of carbide is formed during solidification of the nickel-based superalloy. Previous investigators have reported how to control the morphology of MC carbides for superalloys [8, 11]. The MC carbides may break down to form M_6C and/or $M_{23}C_6$ carbides during high temperature treatments; so a size, morphology and a type of a carbide can be changed during heat treatment of the superalloy.

MAR–M247 is a hafnium-modified superalloy. Hafnium has been known to have a strong effect on carbide morphology in cast alloys, and subsequently on the mechanical properties [12].
Two kinds of MC carbides (precipitated along grain boundaries, as well as coarse and elongated Ta, Hf, W enriched) and fine $M_{23}C_6$ carbides were present in the microstructure of MAR-M247 superalloy after heat treatment in the work of Wawro [13]. The coarse carbides are 2 to 10 μm in diameter and the fine carbides are 0,2 to 0,8 μm in diameter. Ta and Hf enriched MC carbides predominated in the matrix. Chromium–rich $M_{23}C_6$ carbides precipitate preferentially in the periphery of MC carbides and in the (γ + γ') eutectic at grain boundaries. The area fraction of the carbides in MAR–M247 superalloy is about 2.4% [11].

MAR–M247 superalloy is characterized by a low plasticity in creep conditions at moderate temperature with great stresses. That problem is mainly connected with a presence of elongated Chinese script-like MC carbides because they promote initiation and propagation of cracks [7,14].

Quantitative metallography methods have been applied for investigation of carbides occurring in nickel-base superalloys [15-16]. The investigations first of all aim at optimization of chemical content of superalloys. It is not uncommon to have an incomplete description of carbides in superalloys presented in scientific works [7,17-19].

2. Material

The superalloy studied in this work are based on as-cast MAR–M247 nickel-based superalloy after thermal cycling. The chemical composition of the superalloy with 0.15 wt. % C content and several carbide-forming elements is presented in table 1. MAR–M247 is strengthened by γ' precipitates and carbides. There is also solid-solution strengthening of the Ni matrix by the alloying elements. This material is applied mainly to turbine blades and others elements of turbine engines because of its high creep and oxidation resistance.

Table 1. Chemical composition of MAR–M247 superalloy (wt. %)

Cr	Ni	Co	Mo	W	Ta	Ti	Al	Fe	C	Others
8.25	bal.	10.0	0.7	10.0	3.0	1.0	4.5	< 0.5	0.15	0.015 B, 0.05 Zr, 1.5 Hf

3. Experimental

The presence of carbides in polycrystalline MAR–M247 superalloy was confirmed by X-ray diffraction (XRD). The X-ray analysis carried out on the powder of the superalloy indicated the presence not only MC (TiC, WC, TaC), but also small amount of $M_{23}C_6$ ($Cr_{23}C_6$) carbides [20].

The results of X–ray diffraction were verified by studies on the material using an Hitachi HD-2300A scanning transmission electron microscope (STEM) with an accelerating voltage of 200kV. The carbide phases were identified using selected area electron diffraction (SAED).

Chemical composition analyses of the MC carbides conducted using energy dispersive spectrometry (EDS) system attached to SEM developed quantitative chemical analyses of the phases and discovered the occurrence of several variations of these carbides with various contents of carbide forming elements.

The samples of MAR–M247 superalloy for microstructural examinations were ground, polished and chemically etched. The process of sample preparation with applied machine parameters for light microscopy (LM) and scanning electron microscopy (SEM) was described in detail in [15]. Selection of etching methods of primary carbides in MAR–M247 superalloy was discussed in reference [21].

The foils for STEM were mechanically thinned to 0.03-0.05 mm and further electropolished with a solution of 10% perchloric acid and 90% ethanol at -25°C.

4. Results and discussion

Light microscope (LM) was first of all applied to the observation of carbides distributed in the microstructure of the superalloy. The observation of the polished and etched specimens by means of an Olympus GX71 light microscope revealed not only carbides in the microstructure of MAR-M247 superalloy but also γ' phase precipitates and (γ+γ') eutectic (figure 1).

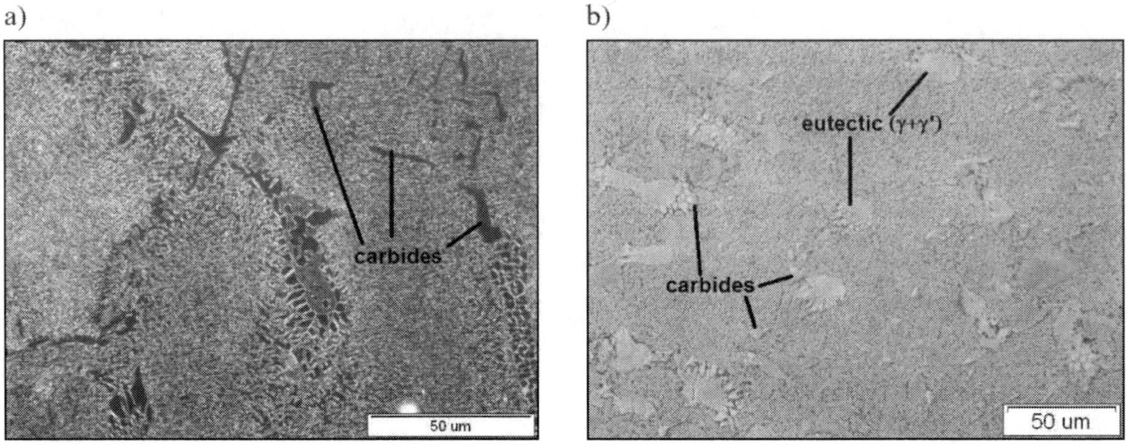

Figure 1. Microstructure of MAR–M247 superalloy. LM, polarized light (a) and DIC (b). The sample etched with solution of HCl and H_2O (a) and 1,5g $CuCl_2$, 33ml H_2O, 33ml HCl, 33ml C_2H_5OH (b).

The light microscopy was not useful for the studies of the carbides in detail; hence, the microstructure was examined using an Hitachi S-4200 SEM as well as Hitachi S-3400N SEM, both equipped with energy-dispersive spectrometer EDS system of Noran.
Observations by means of scanning electron microscope (SEM) obtained more information about the morphology of MC carbides (figure 2) as well as the location of MC and $M_{23}C_6$ carbides (figures 3-4).

The back-scattered electron (BSE) images made it possible to better differentiate grey level differences between the carbides and the matrix. The bright carbides on the images registered using BSE techniques are highly visible against the dark background of the matrix on the unetched sample (figures 4a-d) as well as on the etched sample (figures 4e,f).

On the basis of the SEM studies that were carried out, it has been determined that the microstructure of the examined material consists of γ solid-solution matrix, γ' phase precipitates with a cubic shape, the (γ+γ') eutectic and MC and $M_{23}C_6$ carbides.

The carbides are located at the grain boundaries (figure 4a) and in the interdendritic regions (figures 4a,c,d). Numerous examples of MC carbides occurred within the (γ+γ') eutectic (figures 3b,e). It can be seen that the carbide morphology varied from coarse script-like (figure 4d), fine script-like (figure 4b) to isolated blocky (figures 2d, 4c) or discontinuous particles (figure 4b). Most of the MC carbides have a blocky morphology and are located at the grain boundaries as well as in the interdendritic regions The predominant morphology of the carbides is so-called "Chinese script" with long arms extending to the grain boundaries and primary γ' islands.

Chromium-rich $M_{23}C_6$ carbides occur at the grain boundaries (figure 3d) or outer edges of the primary γ' island (figures 3c, 3f).

The analysis results of the MC carbides in various morphology by EDS are listed in table 2. A compositional fluctuation exists among the MC carbides in the examined superalloy specimens (table 2).

The chemical composition analyses showed that MC carbides in MAR-M247 were complex composition with various contents of carbide forming elements. The presence of nine (not three as described in [22]) types of primary carbides with various contents of carbide forming elements: (Ta,W,Ti,Hf)C – figure 5a; (Ta,W,Hf,Ti) – figure 5b; (Ta,Hf,Ti)C – figure 5c; (Hf,Ti)C – figure 5d; (Ta,Hf,Ti)C – figure 5e; (Hf,Ti)C – figure 5f; (Ta,Ti)C – figure 5g; (Ta,Hf,Ti,W)C – figure 5h and

IOP Conf. Series: Materials Science and Engineering **35** (2012) 012006 doi:10.1088/1757-899X/35/1/012006

(Hf,Zr,Ti)C containing a sulfide in the core – figure 5i was confirmed. In the most cases, the MC carbides contained hafnium.

EDS analyses of the script-like carbides revealed that the script arms and cores contained Ti, Ta, Hf, W and a small amount of Cr. The heads of the MC carbide script structures contained larger amounts of Hf than the cores and arms. Very often, MC arrow-shaped carbides were rich in Hf (figure 5c and table 2). The blocky MC carbides contained significantly higher Hf content than the needle-shaped carbides (figure 5b,d and table 2).

The complex chemical composition of the carbides was confirmed using SEM mapping – figure 6.

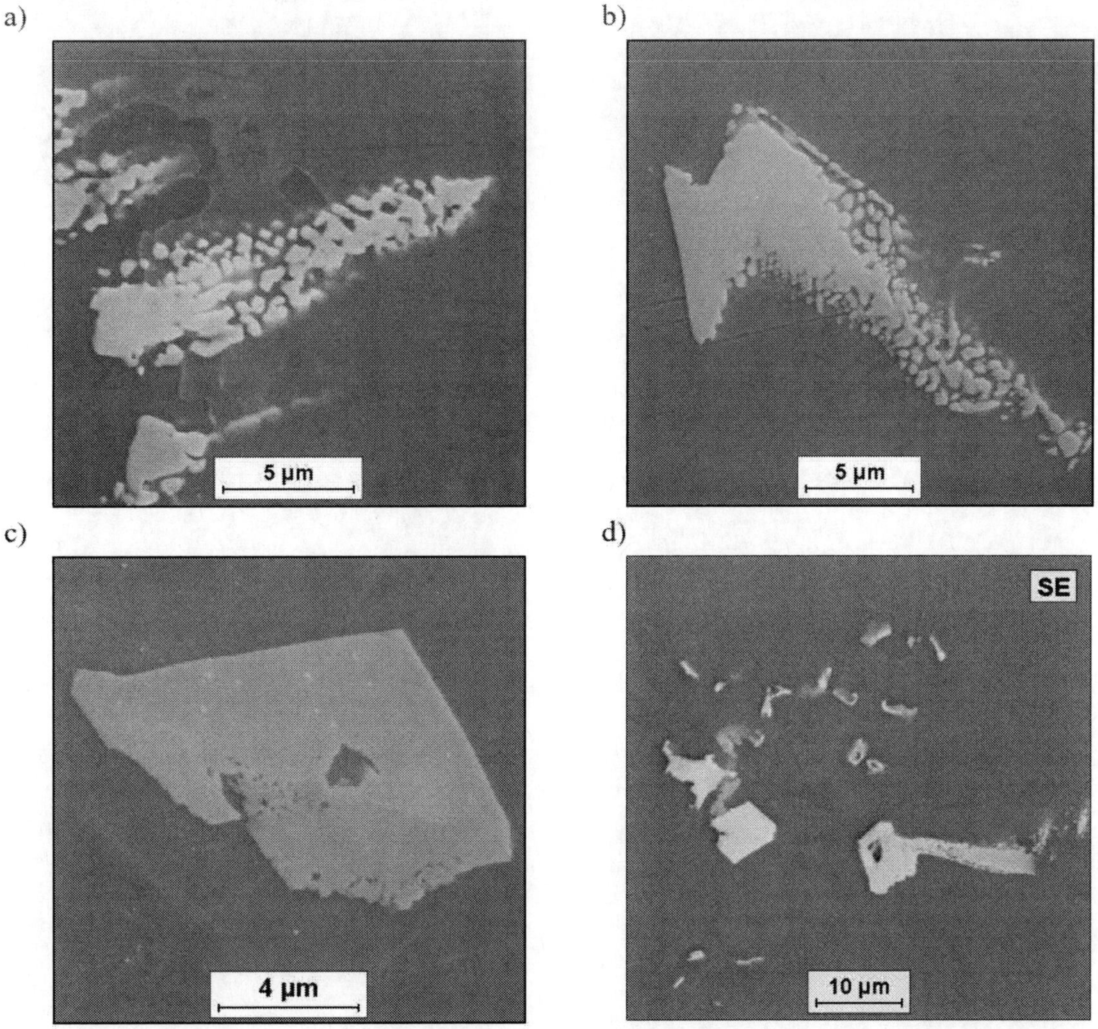

Figure 2. Comparison of carbides' morphology in MAR–M247 superalloy microstructure. SEM, SE images. The sample unetched

In figures 6-9, selected TEM micrographs with selected area electron diffraction (SAED) patterns are presented.

On the basis of the observations performed, one can state that primary MC carbides occur in MAR-M247 nickel-based superalloy as isolated particles as well as in clusters (figure 7). Occasionally, regular-shaped $M_{23}C_6$ carbide were present inside an MC carbide (figure 8). In many cases, $M_{23}C_6$ carbides occurred around MC carbides creating something looking like "lace" (figure 9).

IOP Conf. Series: Materials Science and Engineering **35** (2012) 012006 doi:10.1088/1757-899X/35/1/012006

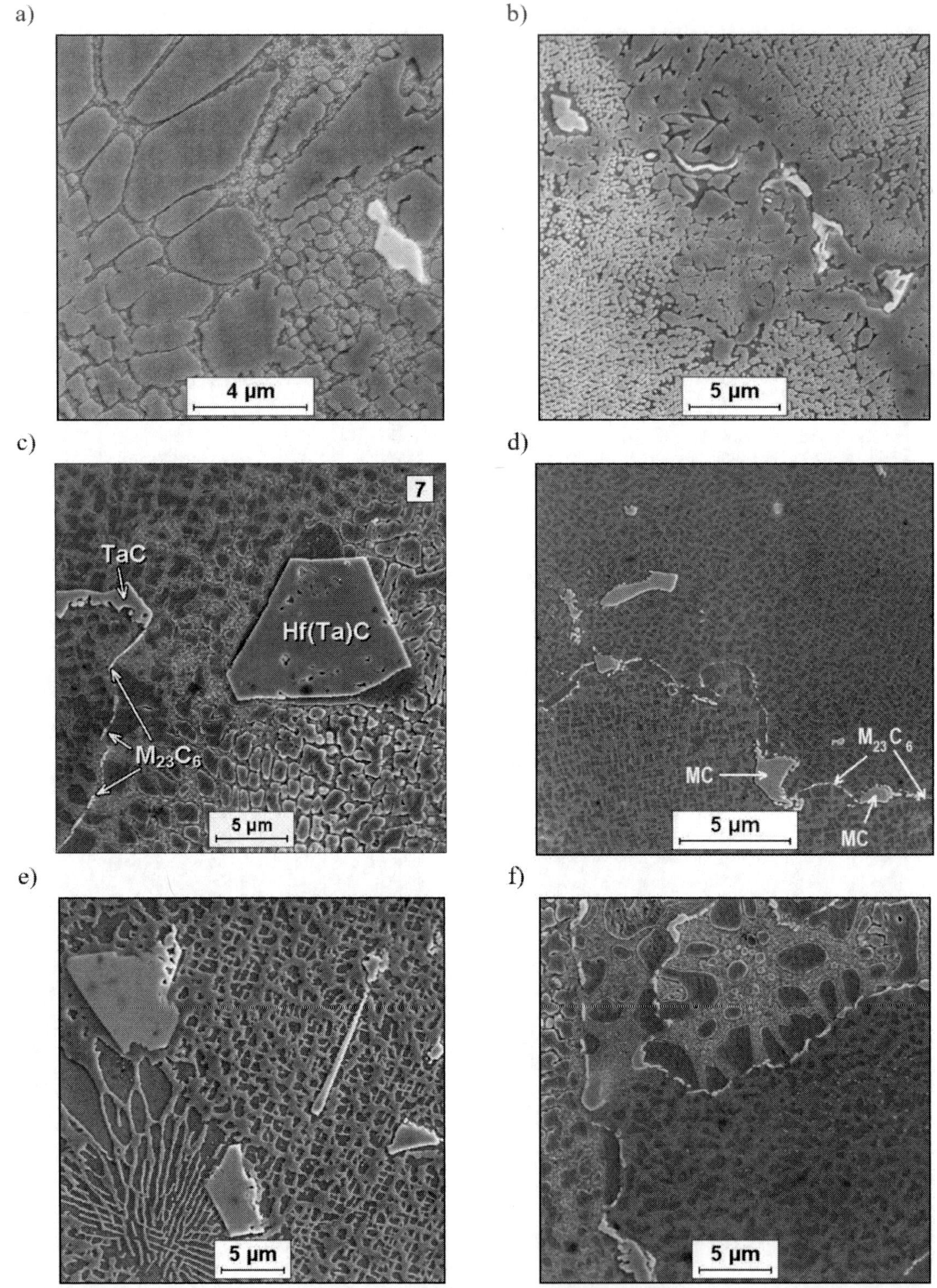

Figure 3. Carbides in MAR–M247 superalloy microstructure. SEM, SE images. The sample was etched with a H_3PO_4 solution (a-d) and with 50 ml of $C_3H_6O_3$ (lactic acid), 30 ml of HNO_3, 2 ml H_2O (e,f).

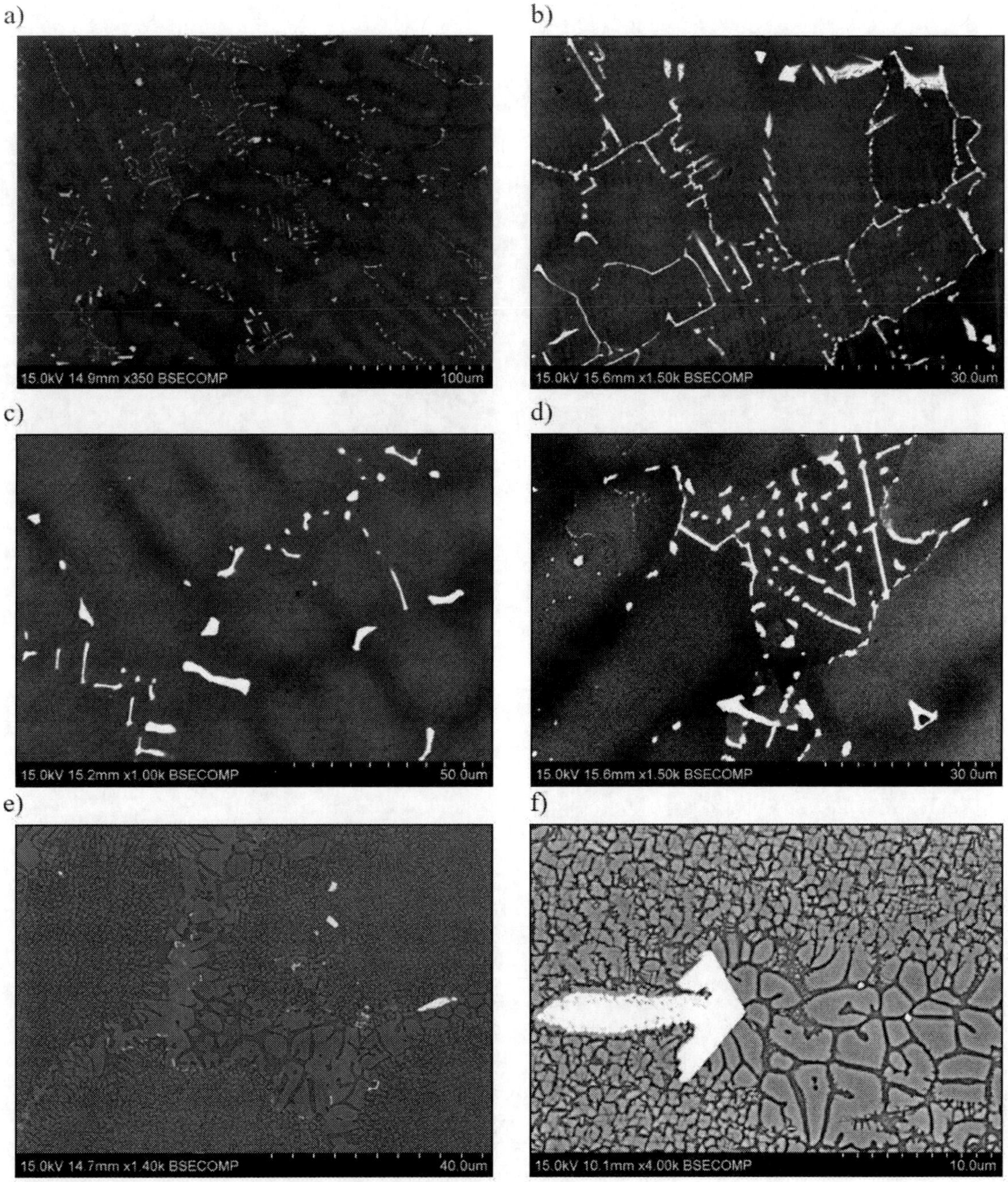

Figure 4. Carbides in MAR–M247 superalloy microstructure. SEM, BSE images. The sample was unetched (a-d) and etched with H_3PO_4 solution (e,f).

Many fine $M_{23}C_6$ carbides were very often located on the γ/γ' interface (figure 10 with carbides between γ' participates) as well as at the MC/γ interface.

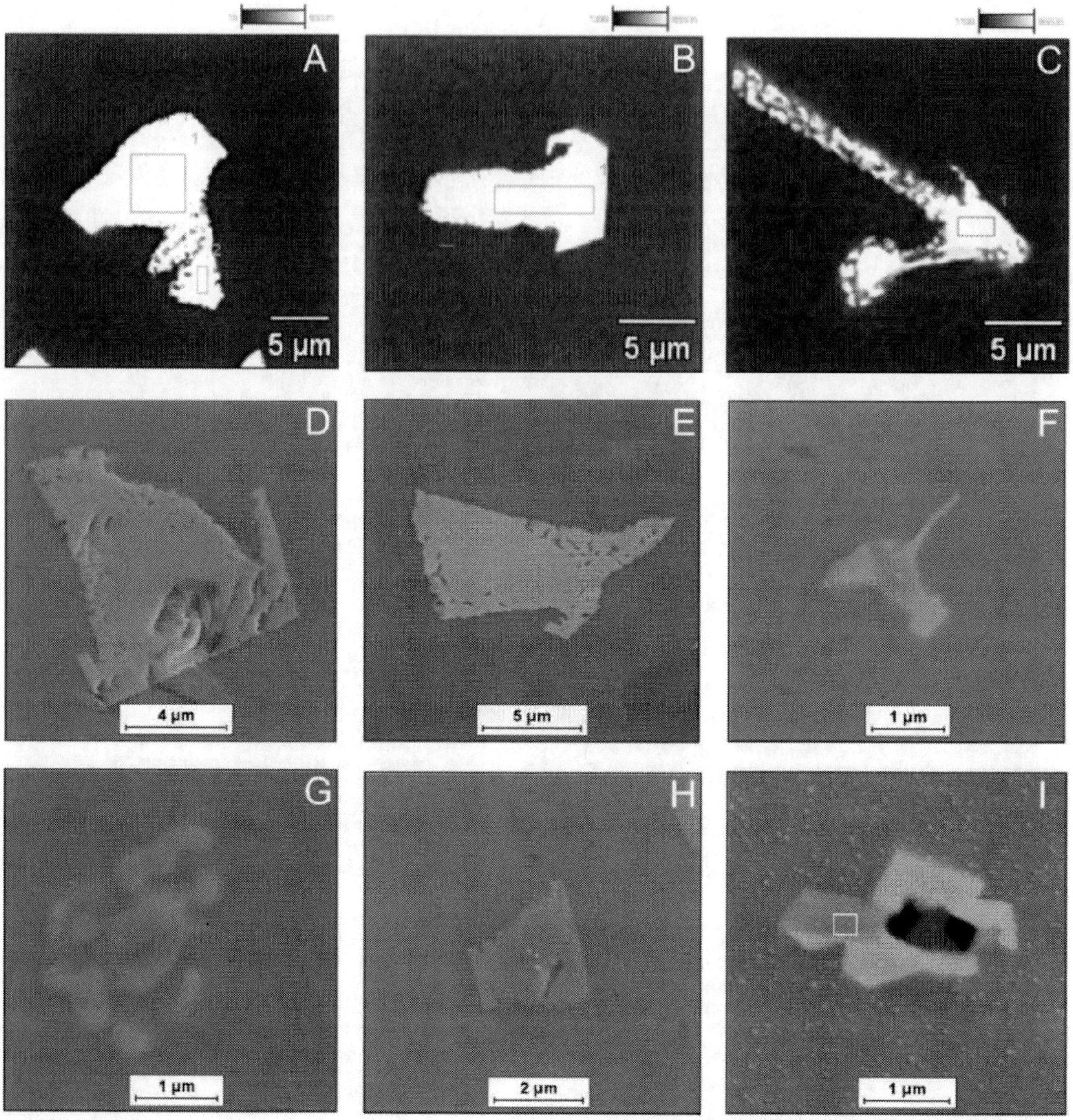

Figure 5. Comparison of MC carbide morphology in the MAR-M247 superalloy. SEM images.

Table 2. Comparison of chemical composition of the MC carbides from figure 5 EDS.

Carbide	Element content															
	Ti		Cr		Co		Ni		Hf		Ta		W		Zr	
	m	a	m	a	m	a	m	a	m	a	m	a	m	a	M	a
A1	6.5	18.7	0.7	1.8	0.80	1.9	4.4	10.4	37.4	28.9	39.9	30.5	10.3	7.8		
A2	8.0	19.9	1.3	2.9	1.50	3.0	11.2	22.7	22.2	14.9	44.4	29.2	11.4	7.4		
B	3.9	11.9	0.7	1.9	1.0	2.7	4.5	11.0	54.2	44.1	26.1	20.9	9.6	7.5		
C	1.4	5.0							91.4	88.0	7.2	7				
D	4.9	1.6					8.7	21.3	86.4	77.1						
E	12.3	34.5							36.5	27.5	51.2	38				
F	6.9	1.6					13.0	17	80.1	81.4						
G	14.8	33.5					12.4	22.9			72.8	43.6				
H	27.3	12.0	5.7	2.7	4	2	9.1	9.8	22.1	14.8	21.1	51.7	10.7	7		
I	1.2	3.9					4.6	11.5	86.4	71.9					7.8	12.7

where m – wt.%, a – at.%

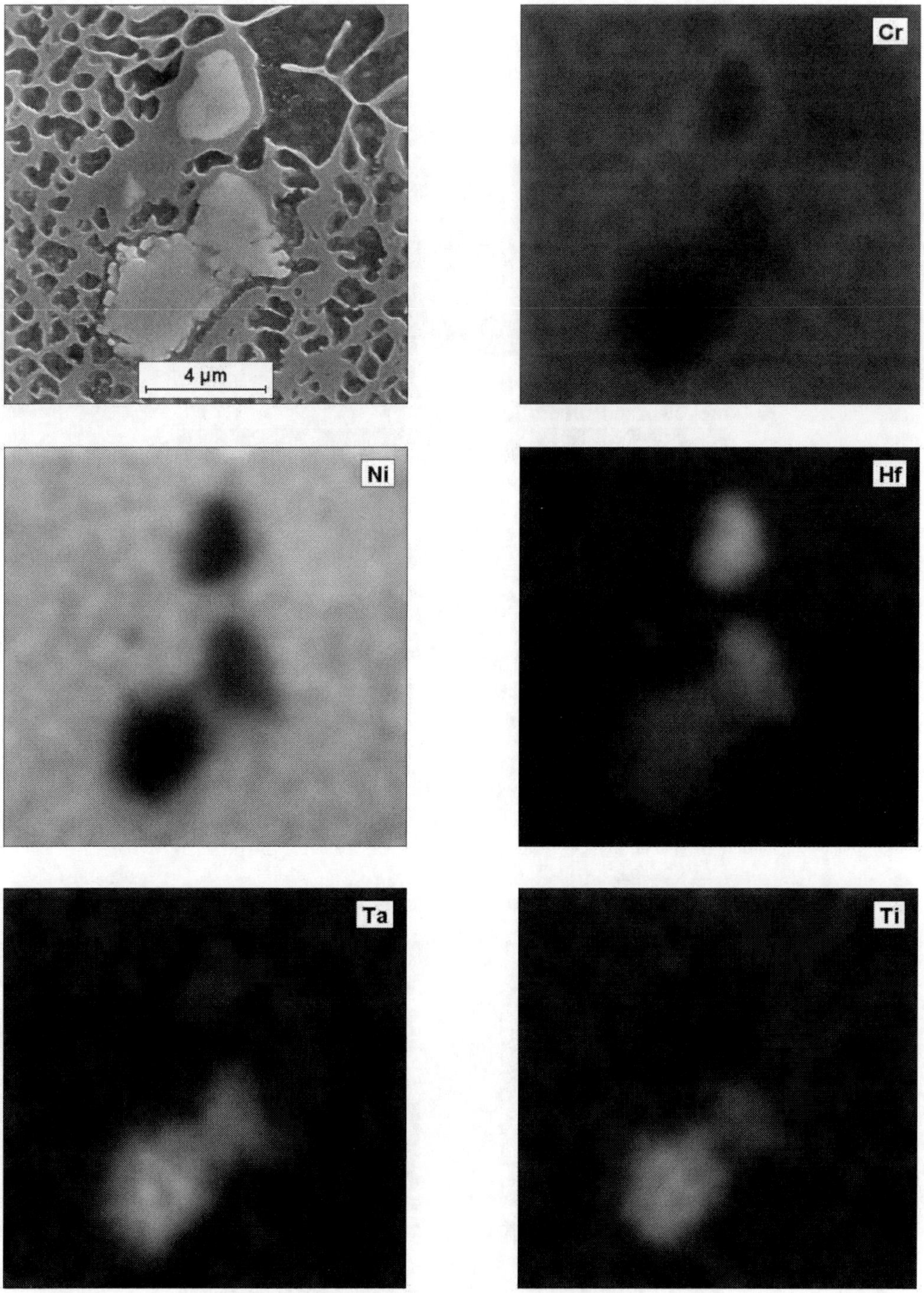

Figure 6. Two MC carbides: HfC and (Hf, Ta, Ti)C and chromium-rich fine $M_{23}C_6$ carbides in MAR-M247 superalloy. SEM, SE image and elemental mapping.

IOP Conf. Series: Materials Science and Engineering **35** (2012) 012006 doi:10.1088/1757-899X/35/1/012006

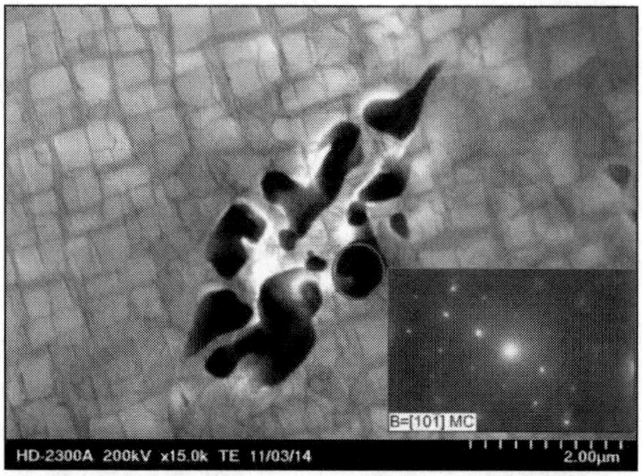

Figure 7. A cluster of MC carbides in MAR-M247 nickel-based superalloy.

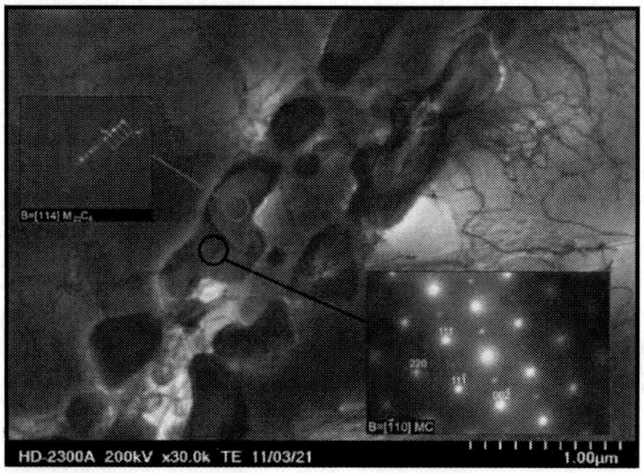

Figure 8. An $M_{23}C_6$ carbide inside an MC carbide in MAR-M247 nickel-based superalloy.

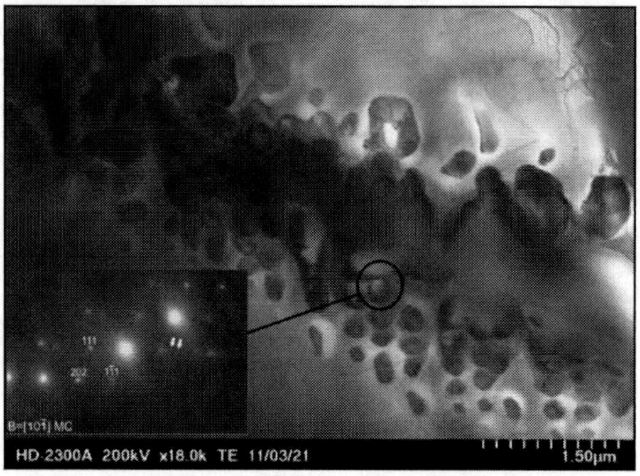

Figure 9. The "lace" of $M_{23}C_6$ carbides around the MC carbide in MAR-M247 nickel-based superalloy. STEM image and SAED pattern (slightly visible reflections close to each other marked with arrows indicate $M_{23}C_6$ carbide).

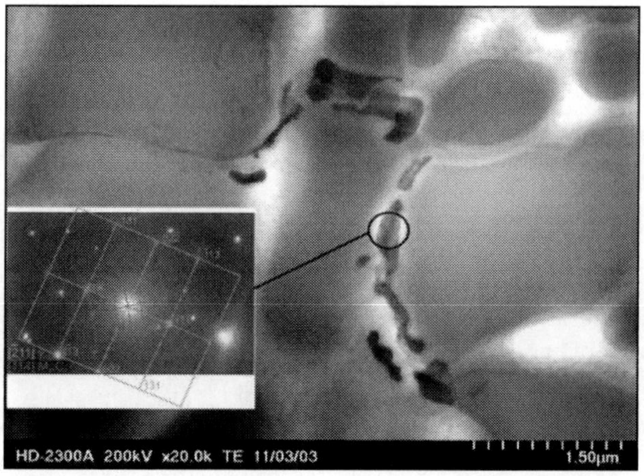

Figure 10. $M_{23}C_6$ carbides between γ' precipitates in MAR-M247 nickel-based superalloy. STEM image and SAED pattern.

4. Conclusion

The microstructure of MAR-M247 nickel-based superalloy is strengthened by γ' precipitates and carbides. The studies performed confirmed the presence of primary MC and secondary $M_{23}C_6$ carbides in an as-cast MAR-M247 superalloy after thermal cycling.

The carbides are located at the grain boundaries and in the interdendritic regions. Numerous MC carbides occur within or at the periphery of the (γ+γ') eutectic. The morphology of MC and $M_{23}C_6$ carbides is very complex. MC carbides are present as discrete blocky precipitates of a diversified shape and size as well as in clusters. The predominant morphology of those carbides visible at low magnifications (approximately up to 1500 times) are "Chinese script-like" structures at interdendritic regions.

The MC carbides in the superalloy examined are characterized by a complex composition with various contents of carbide forming elements. The EDS results prove that several distinct types of MC carbides exist in the investigated MAR-M247 superalloy.

The secondary $M_{23}C_6$ carbides are very fine and exist at the grain boundaries and the outer edges of the primary γ' islands. They often formed a characteristic „lace" around MC carbides and were identified in the vicinity of MC carbides and even inside them. $M_{23}C_6$ carbides also separate from solution at γ/γ' interfaces or at MC/γ interfaces, as well as at the boundaries of matrix grains.

The reaction of MC carbide degeneration ($MC + \gamma \rightarrow M_{23}C_6 + \gamma'$) is well known. It takes place during thermal exposure of the nickel-based superalloy. It is a diffusion-controlled process. The products of the degeneration are $M_{23}C_6$ and γ' on the MC/γ interface. This study confirmed that the process occurs.

Acknowledgements

Financial support of Structural Funds in the Operational Program - Innovative Economy (IE OP) financed from the European Regional Development Fund - Project No POIG.0101.02-00-015/08 is gratefully acknowledged.

References

[1] Donachie M J and Donachie S J 2002 *Superalloys - A Technical Guide* (ASM International, Materials Park, Ohio, USA).

[2] Sims C T, Norman S S and Hagel W C 1987 *Superalloys II* (John Wiley & Sons Inc., New York, USA).
[3] Mitchell A, Cockcroft S L, Schvezov C E, Loquet J N, Fernihough J and Schmalz A J 1996 *High Temperature Materials and Processes* **15** No 1/2 27.
[4] Liu LR, Jin T, Zhao NR, Sun XF, Guan HR, Hu ZQ 2003 *Mater. Sci. Eng.* **A361** 191.
[5] He L Z, Zheng Q, Sun X F, Guan H R, Hu Z Q, Tieu A K, Lu C, and Zhu H T 2005 *Mater. Sci. Eng.* **A397** 297.
[6] Yang J, Zheng Q, Sun X, Guan H and Hu Z 2006 *Mater. Sci. Eng.* **A429** 341.
[7] Chen J, Lee J H, Jo C Y, Choe S J and Lee Y T 1998 *Mater. Sci. Eng.* **A247** 113.
[8] Chen Q Z, Jones C N and Knowles D M 2002 *Scripta Mater.* **47** 669.
[9] He L Z, Zheng Q, Sun X, Hou G, Guan H and Hu Z 2005 *J. Mater. Sci.* **40** 2959.
[10] Yang J, Zheng Q, Sun X, Guan H and Hu Z 2006 *J. Mater. Sci.* **41** 6476.
[11] Bor H Y, Chao C G and Ma C Y 1998 *Scripta Mater.* **38/2** 329.
[12] Kotval P S, Venables J D and Calder R W 1972 Role of hafnium in modifying the microstructure of cast nickel-base superalloys *Metall. Mater. Trans. B* **3/2** 457.
[13] Wawro S W 1982 MC carbide structures in Mar-M247 *NASA Report 167892* (Lewis Research Center, USA).
[14] Saam A and Radavich J F 1982 *Microstructural Sci.* **10** 393.
[15] Szczotok A 2011 *IOP Conference Series: Materials Science and Engineering* **22** 012007.
[16] Belan J 2011 *Mater. Eng.* **18** 121.
[17] Baldan A 1991 *J. Mater. Sci.* **26** 3879.
[18] Yunrong Z and Yulin C 1980 Phase transformations in hafnium-bearing cast nickel-base superalloys *Proceedings of Fourth International Symposium "Superalloys", Champion, Pennsylvania* 465.
[19] Wei C N, Bor H Y, Ma C Y and Lee T S 2003 *Mater. Chem. Phys.* **80** 89.
[20] Szczotok A 2007 *Inżynieria Materiałowa* **3-4** 468.
[21] Szczotok A, Szala J, Cwajna J and Hetmańczyk M 2006 *Mater. Charact.* **56** 348.
[22] Lund C and Radavich J F 1980 Effects of refractory additions on the structure and mechanical properties of a Hf containing nickel base superalloy *Proceedings of Fourth International Symposium "Superalloys", Champion, Pennsylvania* 85.

IOP Conf. Series: Materials Science and Engineering **35** (2012) 012007 doi:10.1088/1757-899X/35/1/012007

Research of the microstructure and precipitation strengthening in a high-temperature Fe–Ni superalloy

K J Ducki
Department of Materials Science, Silesian University of Technology, Krasinskiego 8, 40-019 Katowice, Poland

E-mail: kazimierz.ducki@polsl.pl

Abstract. The paper describes the relationships between the kinetics of precipitation and growth of the intermetallic phase γ' - $Ni_3(Al,Ti)$ and the strengthening magnitude obtained in a high-temperature Fe–Ni superalloy of the A-286 type. In order to accomplish the goal of the study, the author used the LSW coagulation theory and Brown and Ham's conventional analysis of strengthening by ordered particles. The samples were subjected to a solution heat treatment at 980°C/2h/water and then aged at 715, 750 and 780°C, with holding times 0.5-500 h. The heat-treated samples were subjected to structural analyses (TEM, X-ray diffraction) and analyses of mechanical properties (hardness test, static tensile test). Direct measurements on the electron micrographs allowed to calculate the structural parameters of the γ' phase, i.e. mean diameter, volume fraction and mean distance between particles. In accordance with the LSW theory, linear dependencies of changes in mean diameter as a function of ageing time ($t^{1/3}$) were elaborated and the activation energy (E) of the γ' phase coagulation process was determined. The author carried out analyses of strengthening and flow stress ($\Delta\tau_o$) increases as a function of the particle size of the γ' phase and determined the value of the antiphase boundary energy (γ_{apb}) for the analyzed Fe–Ni alloy. It was found that the value of APB energy of phase γ' depended significantly on the alloy ageing temperature. Such a dependence of the quantity γ_{apb} on the ageing temperature for the investigated alloy was explained by an increase in the degree of internal arrangement of phase γ' and by increasing sizes of ordered domains as a result of coalescence.

1. Introduction

High-temperature Fe–Ni alloys that have been precipitation strengthened by intermetallic phases display numerous characteristic properties, i.e. [1-5]: excellent mechanical properties, considerable creep resistance and heat resistance at slightly elevated and high temperatures, excellent corrosion resistance, high ductility at low temperatures, and are non-magnetic. The temperatures at which these alloys are applicable range from the temperature of liquid helium (-269 °C) to 540÷815 °C. This group of modern metallic materials have found an increasingly wide range of applications in conventional and nuclear power industry, in aviation technology, in chemical and petrochemical industries and in cryogenic engineering; additionally, they are used for tools in non-ferrous metal processing.

The principal strengthening phase, which is precipitated in creep resisting Fe–Ni alloys, is the intermetallic phase γ' – $Ni_3(Al,Ti)$. In the initial stage of aging, this phase precipitates in the form of zones and dispersive particles, which are coherent with the matrix. Diffusive processes of growth and coagulation of the phase γ' particles as well as overageing, which is primarily related to the $\gamma' \rightarrow \eta$

(Ni_3Ti) transformation, occur during the subsequent stages of ageing or usage. The process of growth of the phase γ' particles for coagulation controlled with volume diffusion can be described using LSW (Lifshitz-Slyozov-Wagner) theory [6,7]. The LSW coagulation theory proposes that the average diameter of precipitates increases with the cube root of ageing or usage time.

In Fe–Ni alloys that have been precipitation strengthened with the intermetallic phase γ', the intensity of strengthening depends on the following factors [1,8-13]: antiphase boundary (APB) and error energy of the phase γ' arrangement, strength of the matrix (γ) and the γ' phase, cohesion stress γ/γ', volume fraction of the γ' phase, particle size of the γ' phase, indexes of diffusion in γ and γ' phases, and degree of the γ/γ' lattice misfit. Among the above-mentioned factors, cohesion stress occurring between the γ' phase particles and the matrix were for a long time considered the most significant for precipitation strengthening. However, according to numerous authors [1,8-13], the γ' phase is coherent with austenite but cohesion stresses are insignificant and should not be considered the primary factor of the strength increase in this group of alloys.

Application of Brown and Ham's [12] conventional analysis in describing precipitation strengthening enables the determination of the impact of the γ' phase precipitates on the value of critical shear stress or yield point of the aged Fe–Ni alloys. Strengthening by ordered particles of the γ' type is associated with the occurrence of antiphase boundary (APB) on the slip plane of a particle cut by dislocation. The energy of APB per unit area (γ_{APB}) describes a force suppressing the dislocation motion per unit of its length when moving through a particle. According to numerous researchers [1,8-13], the activity of antiphase boundaries within incompletely ordered particles of the γ' phase is the most important mechanism of precipitation strengthening in Fe–Ni alloys.

This study applied LSW coagulation theory and Brown and Ham's conventional analysis to describe the precipitation strengthening by ordered particles of the γ' phase in a high temperature creep resistant Fe–Ni alloy of the A-286 type during heat treatment. The growth index of the γ' phase particles as a function of ageing parameters was identified, the increase in strength and flow stress of the alloy as a function of particle size of the γ' phase was analyzed, and value of APB energy was determined.

2. Material and procedure

The examinations were performed on rolled bars, 16 mm in diameter, of an austenitic alloy designated X5NiCrTi26-15. The chemical composition of the material is given in table 1.

Table 1. Chemical composition of the investigated Fe–Ni superalloy.

Content of an element, wt.%														
C	Si	Mn	P	S	Cr	Ni	Mo	V	W	Ti	Al	Co	B	N
0.05	0.55	1.25	0.026	0.016	14.3	24.5	1.34	0.41	0.10	1.88	0.16	0.08	0.007	0.0062

The samples for testing cut off these rods were subjected to solution heat treatment at 980°C for 2 h and water quenched and then to ageing at temperatures 715, 750 and 780°C at holding time from 0.5 to 500 h.

The structural examinations were carried out using the thin foil technique on a Jeol JEM-2000 FX transmission electron microscope (TEM). The X-ray phase analysis of isolates was carried out with a Philips PW-1140 X-ray diffractometer.

A quantitative analysis of secondary γ' phase particles, which phase precipitated in the alloy during ageing, was performed on TEM images in bright field mode. Based on direct counting, basic stereological parameters of the γ' phase were determined from binary images, i.e.: the mean diameter ($\overline{D}$), the volume fraction particles (V_V) and the mean distance between particles (l_a).

The mean diameter ($\overline{D}$) of the particles was evaluated with help of the relation given by Czyrska-Filemonowicz et al. [14] and Wasilkowska et al. [15]:

$$\overline{D} = \frac{\overline{d}t}{t - \overline{d} + (\pi A_A / L_A)} \tag{1}$$

where: $\overline{d}$ – the mean diameter of the circles in the projected plane (nm), t – the foil thickness (nm), A_A – the projection area, L_A – the perimeter density (1/nm).

The volume fraction (V_V) of γ' particles was calculated using the formula provided by Dubiel et al. [16]:

$$V_V = \frac{\pi}{6} \frac{\sum N_i d_i^3}{A(t + \overline{d})} \tag{2}$$

where: N_i – the number of particles with diameter d_i, A – the total projection area (nm^2).

The mean distance between particles (l_d) was estimated using the equation given by Schröder and Arzt [17]:

$$l_d = \overline{d}\sqrt{\frac{\pi}{6V_V}\left(1 + \frac{s^2}{\overline{d}^2}\right)} - \frac{\pi}{4}\overline{d} \tag{3}$$

where: s – the standard deviation of the particle diameter distribution.

The analysis of the γ' phase growth step was made on the basis of the LSW theory for coagulation on volume diffusion controll according to the relationship provided by Kusabiraki et al. [18]:

$$\overline{d}^3 - \overline{d}_0^3 = \frac{64\sigma D C_e V_m^2}{9RT} = K't \tag{4}$$

where: $\overline{d}$ and $\overline{d}_0$ – average diameter of precipitates at time t and $t = 0$ respectively, σ – interfacial energy between precipitates and matrix, D – diffusion coefficient of the solute atom in the matrix, C_e – concentration of solute atoms in the matrix in equilibrium with a particle of an infinite size, V_m – molar volume of precipitation phase, R – gas constant, T – absolute temperature, K' – growth rate constant.

The hardness measurements of samples made by Vickers method at 98 N load were performed using a Zwick hardness tester coupled with a PC class computer. The static tensile test of samples were performed on MTS testing machine with extensometric measurement of elongation.

Flow stress increments ($\Delta\tau_o$) for aged Fe–Ni alloy were plotted as a function of square root of γ' particle size in accord with the Brown and Ham [12,13] conventional analysis for ordered-particle strengthening:

$$\Delta\tau_O = \frac{1}{2b} \cdot \left(\frac{\overline{D}f}{T}\right)^{1/2} \cdot (\gamma_{apb})^{3/2} - \frac{f}{2b}(\gamma_{apb}) \tag{5}$$

where: $\overline{D}$ – particle diameter, f – particle volume fraction, b – matrix Burgers vector, T – dislocation line tension, γ_{apb} – anti-phase boundary energy.

3. Results and discusion

The examined Fe–Ni alloy, after solution heat treatment (980°C/2h/water), has the structure of twinned austenite with increased dislocation density and a small amount (ca. 0.3 %) of undissolved particles (figure 1). The following titanium compounds occurred in the phase composition of undissolved precipitates: carbide TiC, nitride TiN, carbonitride Ti(C,N), carbosulfide $Ti_4C_2S_2$, Laves phase Ni_2Si and boride MoB [19].

Application of 1 × step ageing at 715, 750, and 780°C after heat solution treatment and holding times ranging from 0.5 to 500 h results in the processes of precipitation of intermetallic phases of the

IOP Conf. Series: Materials Science and Engineering **35** (2012) 012007 doi:10.1088/1757-899X/35/1/012007

γ' – $Ni_3(Al,Ti)$, η (Ni_3Ti), G ($Ni_{16}Ti_6Si_7$), β (NiTi), and σ ($Cr_{0,46}Mo_{0,40}Si_{0,14}$) types and of $M_{23}C_6$ carbide in the Fe–Ni alloy [19]. The intermetallic γ' phase was the primary phase precipitated during alloy ageing. This phase, in the initial stages of ageing at 715°C, precipitated homogeneously in the form of zones and dispersive particles coherent with the matrix, which, according to Pickering [11], is related to spinodal decomposition of supersaturated austenite (figure 2).

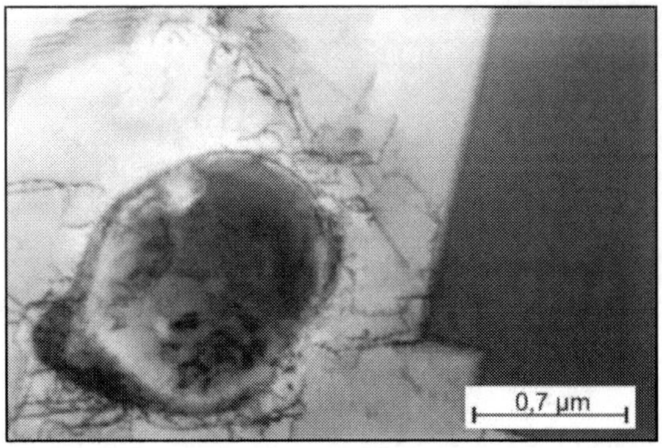

Figure 1. Alloy microstructure after solution heat treatment at 980°C/2h/w. Twinned austenite with an increased dislocation density and an undissolved TiC carbide.

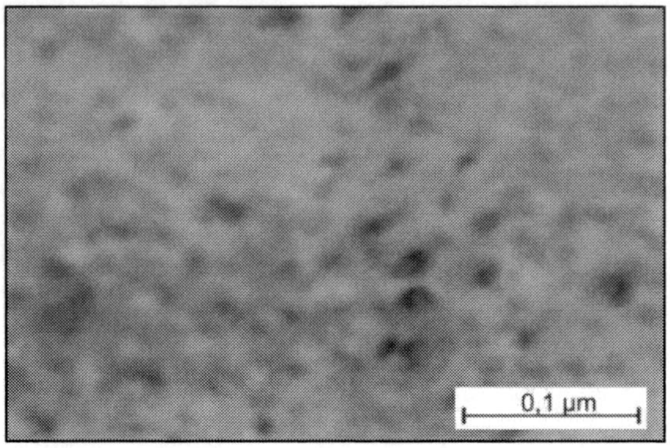

Figure 2. Alloy microstructure after solution heat treatment and ageing at 715°C/4h. Coherent zones and ordered precipitates γ' in the austenite region.

Intensive precipitation of dispersive and spheroidal γ' phase particles characterized by strong cohesion with austenite was observed in the alloy matrix at longer ageing times (figure 3).

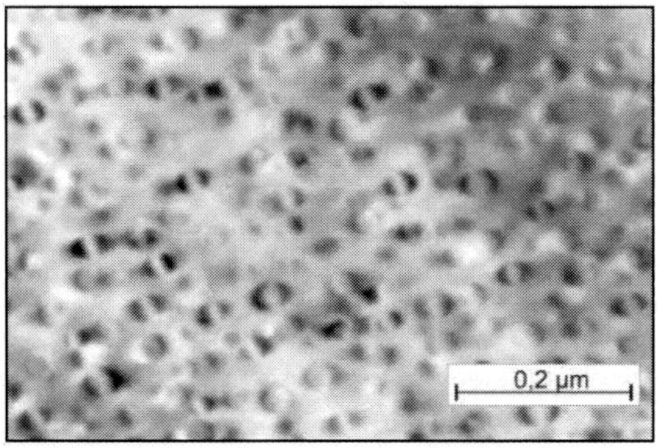

Figure 3. Alloy microstructure after solution heat treatment and ageing at 715°C/150h. Coherent spherical particles of phase γ' in the austenite.

Increase in the aging temperature to 750°C intensified the diffusive processes of growth and coagulation of the γ' phase particles, which still retain a spheroidal shape and cohesion with austenite (figure 4). The structure of alloy shows also first effects of overageing into form of transcrystalline Widmanstätten plates and cellular precipitates of η (Ni_3Ti) phase (figure 5).

At the highest ageing temperature of 780°C, a clear effect of overageing, associated with coagulation of the γ' phase particles, loss of cohesion with austenite and occurrences of arrangement errors inside the precipitates were observed in the alloy structure (figure 6).

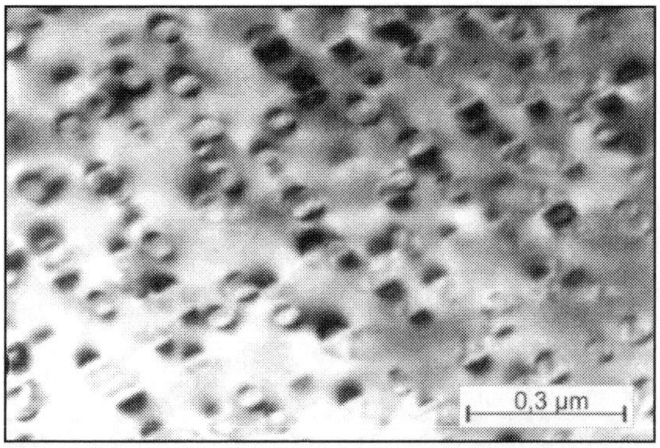

Figure 4. Alloy microstructure after solution heat treatment and ageing at 750°C/300h. Enlarged spheroidal particles of phase γ' coherent with austenite.

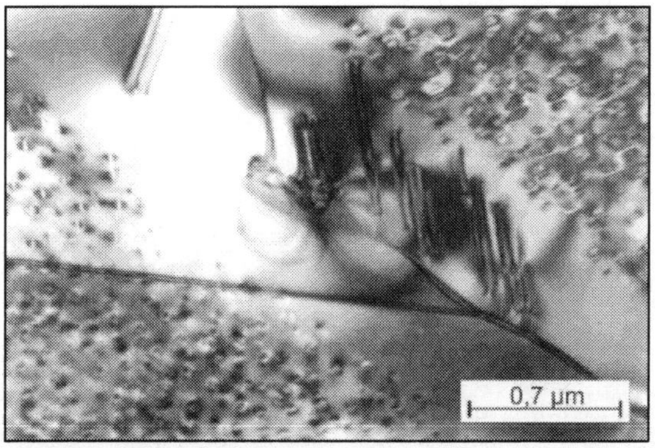

Figure 5. Alloy microstructure after solution heat treatment and ageing at 750°C/400h. Cellular precipitates of η phase in the region near the boundary.

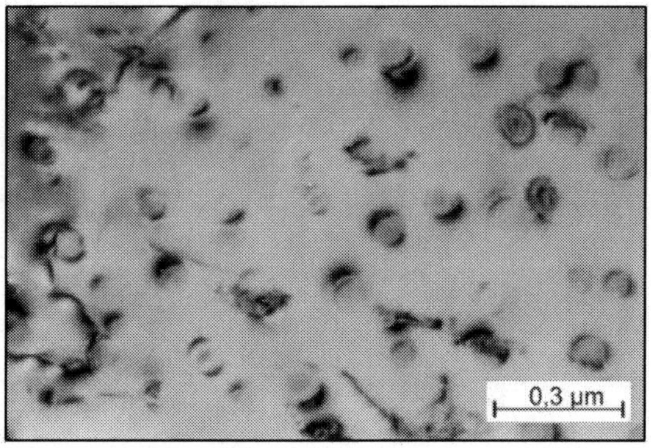

Figure 6. Alloy microstructure after solution heat treatment and ageing at 780°C/500h. Coagulated particles of phase γ' and dislocations in the matrix.

The following compounds were revealed in the phase composition of extracted precipitates: carbide TiC, carbonitride Ti(C,N), nitride TiN, carbosulfide $Ti_4C_2S_2$, boride MoB, and Laves phase Ni_2Si [19].

Quantitative analysis of particles of the γ' phase precipitated in the alloy during ageing was conducted at the ageing temperatures of 715, 750, and 780 °C and holding times ranging from 4 to 500 h. For ageing times below 4 h, only coherent zones occurred in the alloy matrix and quantitative analysis of the γ' phase was not conducted. Table 2 presents results of the quantitative analysis of the γ' phase obtained for selected ageing times, i.e. 4, 150, 300, and 500 h. It is evident from the presented data that at an ageing temperature of 715 °C and holding times of 4-500 h the mean diameter of particles ($\overline{D}$) increased from 7.5 to 28 nm, while the volume fraction changed from 6 to 11.1 %.

For the higher ageing temperature of 750 °C and the same holding times, the mean particle diameter increased from 10.6 to 39.5 nm and the volume fraction from 7.1 to 10.1 %. At the highest ageing temperature of 780 °C and analogical holding times, the mean particle diameter reached the greatest values, ranging from 13.1 to 58.4 nm, with the constant range of volume fraction about 7.7÷8.1 %.

The kinetics of growth and coagulation of the γ' phase was analyzed based on LSW theory according to equation (4). The relationship between the mean diameter ($\overline{D}$) of the γ' phase particles and ageing time raised to the one third power ($t^{1/3}$) at the aging temperatures of 715, 750, and 780 °C is presented in figure 7. Linear regressions were obtained for all analyzed temperatures, which

IOP Conf. Series: Materials Science and Engineering **35** (2012) 012007 doi:10.1088/1757-899X/35/1/012007

demonstrates that coagulation of the γ' phase particles occurs according to LSW theory and is controlled by volume diffusion. The regression slopes increase with rising temperature of ageing, which can be explained by an increase in rates of particle coagulation and growth. In the case of the shortest analyzed ageing times, i.e. 0.5 and 2 h, the mean values of diameters ($\overline{D}$) of the γ' phase precipitates were obtained through extrapolation.

Table 2. Stereological parameters of the γ' particles in the tested alloy.

Parameter	Time of ageing (h) at 715°C				Time of ageing (h) at 750°C				Time of ageing (h) at 780°C			
	4	150	300	500	4	150	300	500	4	150	300	500
A_A (%)	10.2	14.3	14.8	16.4	11.5	15.8	16.6	15.6	9.5	13.5	13.1	11.5
$\bar{d}$ (nm)	7.4	15.7	19.6	24.6	9.2	24.4	27.0	32.1	12.7	41.2	45.0	51.2
$\overline{D}$ (nm)	7.5	17.0	22.5	28.3	10.6	28.8	32.1	39.5	13.1	45.7	52.2	58.4
V_V (%)	6.0	10.3	10.4	11.1	7.1	8.6	9.4	10.1	8.1	7.7	7.8	8.1
l_d (nm)	144	117	121	114	135	141	123	132	153	105	103	124

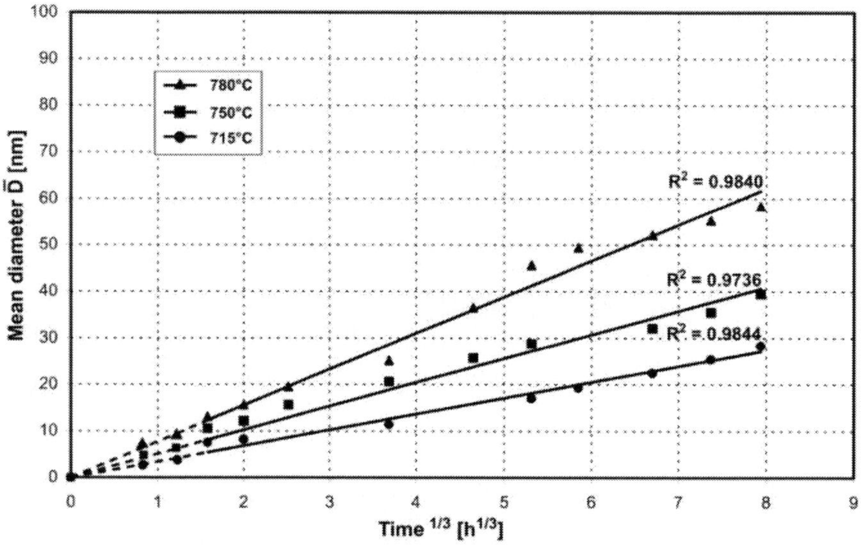

Figure 7. Dependence of the average diameter of phase γ' particles in the Fe–Ni alloy on the ageing time and temperature.

The activation energy of the γ' phase coagulation process in the examined Fe–Ni alloy was determined from the Arrhenius straight line slope (figure 8). The obtained value (E = 297 kJ mol^{-1}) was close to the value of activation energy of phase γ' (245÷298 kJ mol^{-1}), estimated in selected Fe–Ni and Ni-base superalloys [18]. In addition, the determined activation energy value of phase γ' coagulation was similar to the value of activation energy for diffusion of solute elements in austenite (γ) which are considered to contribute to the growth of γ' phase, i.e. Al, Ti, Ni, Fe, Cr and Mo. The value of E obtained for the tested Fe–Ni alloy showed that the growth of γ' phase was controlled by diffusion process of solution elements, which is conformable with theory LSW.

Precipitation strengthening of the Fe–Ni alloy during ageing was described based on results from the structural analysis and investigation of mechanical properties. The initial analysis included plotting graphs of increases in hardness (*ΔHV*), yield strength (*ΔP.S.*), and tensile strength (*ΔT.S.*) as a function of change in the γ' phase particle size, i.e. the mean diameter ($\overline{D}$). The obtained increases in hardness (*ΔHV*), yield strength (*ΔP.S.*) and strength (*ΔT.S.*) as a function of the mean diameter ($\overline{D}$) of the γ' phase particles are presented in figures 9-11. It is evident from the plotted curves that the level of strengthening obtained during ageing was highly correlated with the average diameter of the γ' precipitates and the ageing temperature. The greatest increases in hardness (*ΔHV* = 136÷145) and strength (*ΔP.S.* = 379÷415 MPa, *ΔT.S.* = 391÷416 MPa) were obtained for the ageing temperature of 715 °C and particle size of 7÷11 nm. Slightly less significant increases in hardness (*ΔHV* = 104÷120) and strength (*ΔP.S.* = 288÷311 MPa, *ΔT.S.* = 297÷314 MPa) were noted for the ageing temperature of 750 °C and somewhat larger diameters of the γ' phase precipitates, 7÷12 nm. The least significant increases in hardness (*ΔHV* = 91÷101) and strength (*ΔP.S.* = 236÷272 MPa, *ΔT.S.* = 225÷252 MPa) occurred for the highest ageing temperature, 780 °C, and slightly larger diameters of the γ' phase precipitates ranging from 9 to13 nm.

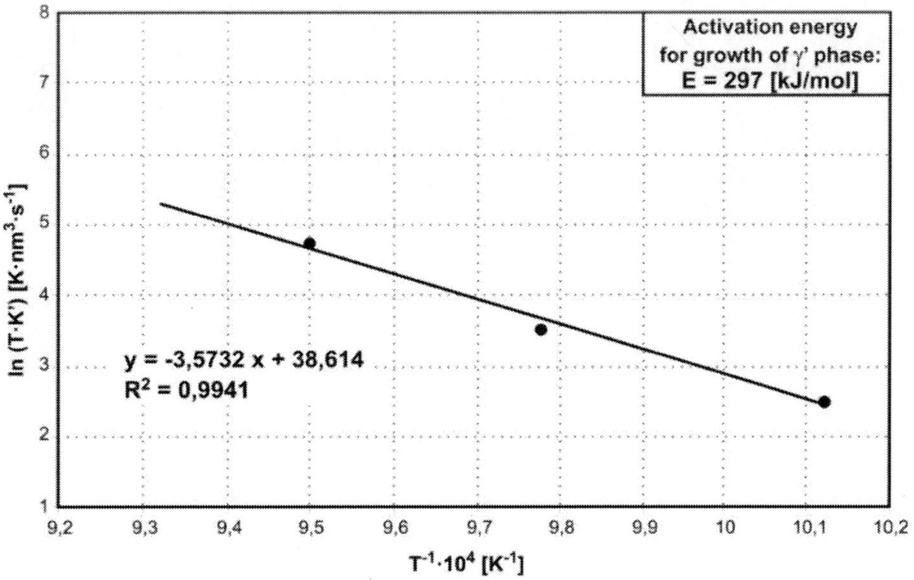

Figure 8. The plot for determination of the activation energy for coagulation process of phase γ' in the Fe–Ni alloy.

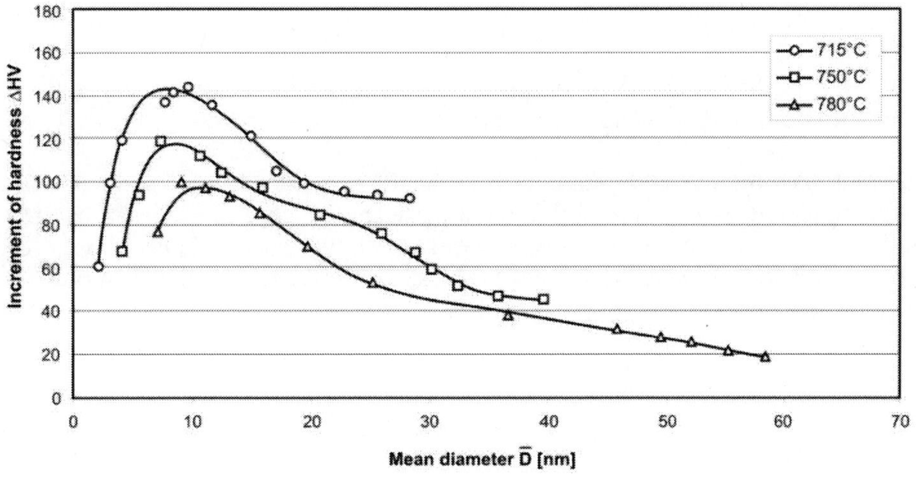

Figure 9. Dependence of hardness increment on the average particles diameter of phase γ' and ageing temperature in the Fe–Ni alloy.

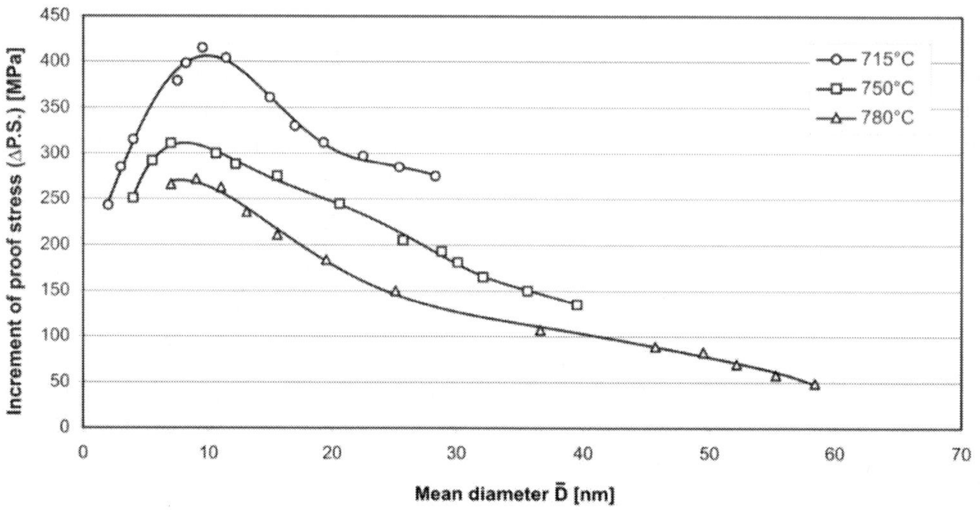

Figure 10. Dependence of proof stress increment on the average particles diameter of phase γ' and ageing temperature in the Fe–Ni alloy.

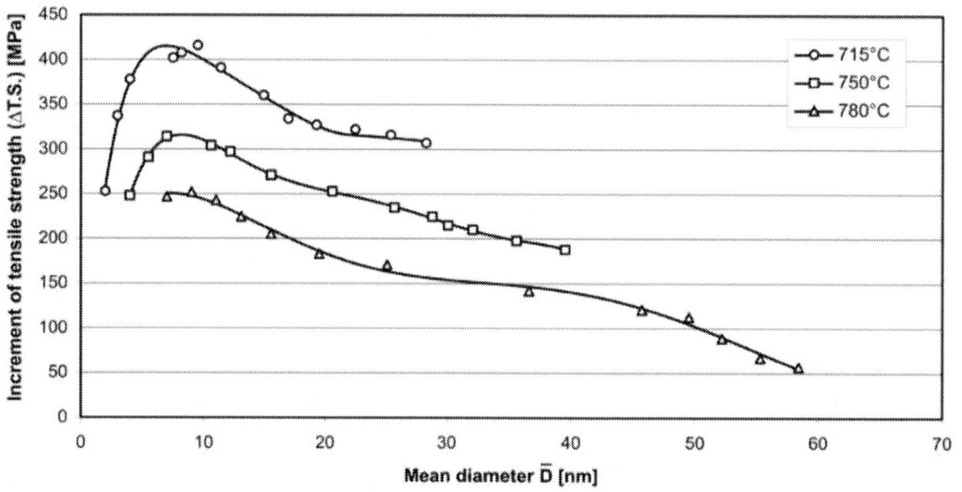

Figure 11. Dependence of tensile strength increment on the average particles diameter of phase γ' and ageing temperature in the Fe–Ni alloy.

The second stage of the alloy strengthening analysis included application of the conventional theory of precipitation strengthening by ordered particles, which was proposed by Brown and Ham [12]. The increase in flow stress ($\Delta\tau_o$) for the tested alloy was analyzed according to equation (5) of this article. In order to determine the values of flow stress for the tested alloy, it was necessary to solve for $\Delta\tau_o$ from the equation [12,13,20]:

$$\Delta\tau_O = \frac{\Delta\sigma_a}{M} \tag{6}$$

where: $\Delta\sigma_a$ – increase in alloy yield point during ageing, M – Taylor index (equals 3.067 for FCC structures).

The resultant values of the flow stress increase ($\Delta\tau_o$) as a function of the mean diameter ($\bar{D}$) of the γ' phase precipitates are presented in figure 12. It is evident from these relationships that the highest increases in flow stress ($\Delta\tau_o$ = 124÷135 MPa) were observed for the ageing temperature of 715 °C and the heated equivalent diameter ($\bar{D}$) of the γ' phase particles of ca. 7÷11 nm. Further increase in the particle diameters from 15 to 23 nm resulted in a rapid decrease in the strengthening level during ageing. For a higher ageing temperature (750°C), the greatest increases in flow stress ($\Delta\tau_o$ = 94÷101 MPa) were obtained for the particle sizes of $\bar{D}$ = 5÷12 nm. Further increase in the particle diameter was accompanied by a continuous decline of strength increase during ageing. For an ageing temperature of 780°C, the maximal increase of flow stress ($\Delta\tau_o$ = 86÷89 MPa) occurred for the particle size of 7÷11 nm. Further increase in the γ' precipitate sizes from 13 to 58 nm resulted in a continuous decrease of the flow stress increase of the tested alloy.

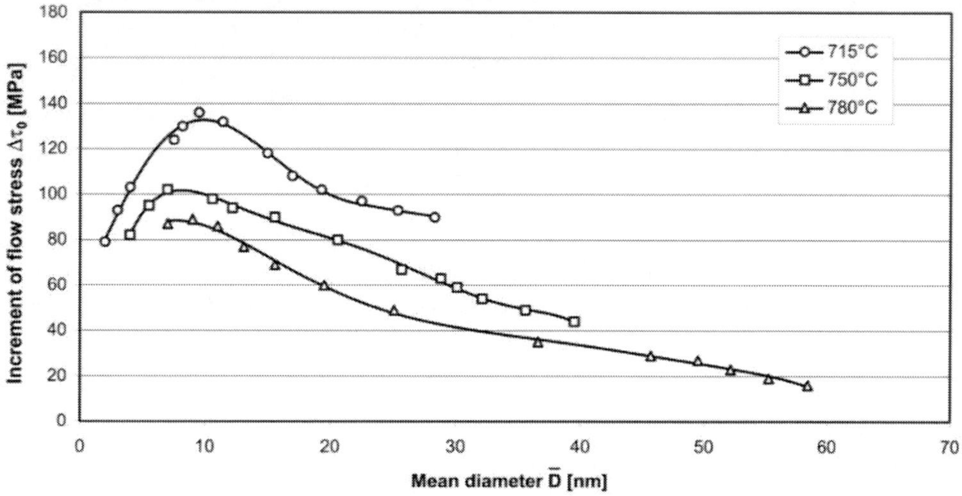

Figure 12. Dependence of critical shearing stress increment on the average particles diameter of phase γ' and ageing temperature in the Fe–Ni alloy.

Based on the results, it can be concluded that the greatest precipitation strengthening in the tested Fe–Ni alloy occurs in the early stages of ageing in the temperature range of 715÷780 °C. For this initial phase of ageing at temperatures of 715, 750 and 780°C, the straight lines were approximated from the growth of critical stress ($\Delta\tau_0$) as a function of the average diameter of particles ($\bar{D}^{1/2}$) of phase γ' (figure 13). Using equation (5) and the inclination of the obtained straight lines, the antiphase boundary energy (γ_{apb}) was determined for the examined range of ageing temperature.

It was fund that value γ_{apb} depended significantly on the ageing temperature, with the highest value (γ_{apb} = 227 mJ m^{-2}) having been obtained for the temperature of 715 °C. An increase in the ageing temperature to 750°C and 780°C causes a decrease of the antiphase boundary energy to 193 mJ m^{-2} and 165 mJ m^{-2}, respectively. Such a dependence of the antiphase boundary energy (γ_{apb}) on the ageing temperature can be explained by the increasing degree of internal arrangement of phase γ' and by the increasing size of ordered domains [11,13]. The phase γ' which precipitates at ageing temperature of 715 °C is a low ordered phase and distinguishes itself with small sizes of domains. Therefore, a higher critical shearing stress ($\Delta\tau_0$) is necessary for such particles to be intersected by the shifting dislocations. The increasing of the alloy ageing temperature to 750 °C and 780 °C leads to increasing the internal arrangement of phase γ'. At the same time, ordered domains grow as a result of coalescence, reducing thereby the stress of "intersecting" the particles, the final effect of which is a visible decrease of precipitation strengthening.

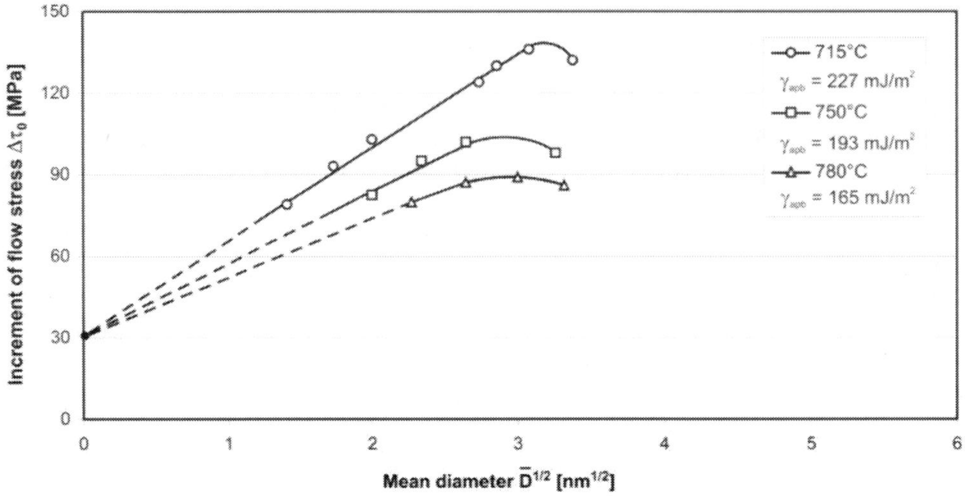

Figure 13. Dependence of the critical growth of shearing stress from the square root of the average diameter of phase γ' particles in the Fe–Ni alloy during the early stages of ageing.

4. Conclusion

The conducted research and data analysis demonstrated a significant impact of heat solution treatment and extended ageing at increased temperatures on the precipitation process, structure, and strengthening of a high temperature alloy of the A-286 type. After the 980°C/2h/water heat solution treatment, the tested Fe–Ni alloy had a structure of twinned austenite with a small amount (ca. 0.3 %) of undissolved precipitates of titanium compounds, i.e. carbide TiC, carbonitride Ti(C,N), carbosulfide $Ti_4C_2S_2$, Laves phase Ni_2Si, and boride MoB.

Application of 1 × step ageing at 715, 750, and 780 °C after heat solution treatment and holding times ranging from 0.5 to 500 h results in processes of precipitation of intermetallic phases of the γ' – $Ni_3(Al,Ti)$, η (Ni_3Ti), β (NiTi), G ($Ni_{16}Ti_6Si_7$) and σ ($Cr_{0,46}Mo_{0,40}Si_{0,14}$) types and of $M_{23}C_6$ carbide in the Fe–Ni alloy. The intermetallic γ' phase was the primary phase precipitated during the ageing of the tested Fe–Ni alloy. Three characteristic stages can be distinguished in the process of γ' phase precipitation: coherent zones, coherent spheroidal particles ($\overline{D}$ = 7÷40 nm) and coagulated spheroidal particles ($\overline{D}$ = 45÷58 nm). It was concluded that the mean diameter ($\overline{D}$) of the γ' phase precipitates increases as a function of the cube root of the ageing time ($t^{1/3}$), which is in conformity with predictions based on LSW theory for coagulation controlled by volume diffusion. The determined activation energy value of phase γ' coagulation was E = 297 kJ mol^{-1} and it was similar to the value of activation energy for diffusion of elements forming phase γ'.

The processes of precipitation, growth and coagulation of intermetallic phases and carbides occurring in the Fe–Ni alloy during aging had a significant impact on the alloy's mechanical properties. The greatest strengthening of the alloy was obtained at an aging temperature of 715 °C. An increase in the aging temperature to 750 and 780 °C was accompanied by clear effects of alloy overaging, which resulted in a decline in hardness and mechanical properties and a slight improvement of plastic properties. It was observed that the level of strengthening obtained in the tested alloy significantly correlated with the γ' phase particle size and ageing temperature. The greatest strengthening of the alloy occurred in initial stages of aging for a mean diameter ($\overline{D}$) of the γ' phase particles ranging from 7 to 11 nm. A decrease in strengthening was observed for greater particle diameters, which is characteristic for the overageing process occurring in alloys that have been precipitation strengthened. Declining strengthening of the alloy with rising ageing temperature can be explained by lower density of the γ' phase precipitates caused by lower nucleation frequency and progress of the $\gamma' \rightarrow \eta$ phase transition.

I was shown that the magnitude of precipitation strengthening in the Fe–Ni alloy in the early stages of ageing in the temperature range of 715÷780 °C may be described by using equations of the conventional theory of strengthening with ordered particles. It was found that the value of APB energy of phase γ' depended significantly on the alloy ageing temperature. The highest value of the antiphase boundary energy (γ_{apb} = 227 mJ m^{-2}) was obtained for the alloy ageing temperature of 715 °C. Increasing of the alloy ageing temperature to 750 °C and 780 °C caused a decrease of the antiphase boundary energy to 193 mJ m^{-2} and 165 mJ m^{-2}, respectively. Such a dependence of the quantity γ_{apb} on the ageing temperature for the investigated alloy was explained by an increase in the degree of internal arrangement of phase γ' and by the increasing sizes of ordered domains as a result of coalescence.

References

[1] Sims Ch T, Stoloff N S and Hagel W C 1987 The Superalloys II, edited by A. Wiley Witescience Publications, New York.
[2] Stoloff N S 2005 in ASM Handbook, Properties and Selection: Irons, Steels and High Performance Alloys, edited by ASM International **1** 1478.
[3] Rohrbach K P 1995 *Advanced Materials & Processes* **10** 37.
[4] Sczerzenie F and Maurer G E 1987 *Mater. Sci. Technol.* **3** 733.
[5] Härkegård G and Guédou J Y 1998 in Proc. of the 6th Liège Conference: Materials for Advanced Power Engineering Part II 913.
[6] Lifshitz I M and Slyozow V V 1961 *J. Phys. Chem. Solids* **19** 35.
[7] Wagner C 1961 *Zeitschrift für Elektrochemie* **65** 581 (in German).
[8] Martin J W 1980 Micromechanisms in particle-hardened alloys, edited by Cambridge University Press, Cambridge.
[9] Gleiter H 1982 in Proc. of the 6th International Conference: Strength of Metals and Alloys, Melbourne **3** 1009.
[10] Ardell A J 1985 *Metallurg. Trans. AIME* **16A** 2131.
[11] Pickering F B 1975 in Proc. of Conference: The Metallurgical Evolution of Stainless Steels, edited by The Metals Society London 391.
[12] Brown L M and Ham R K 1971 Strengthening Methods in Crystals, Applied Science Publishers Ltd, London.
[13] Thompson A W and Brooks J A 1982 *Acta Metallurg.* **30** 2197.
[14] Czyrska-Filemonowicz A, Dubiel B and Wiencek K 1998 *Acta Stereolog.* **17** 225.
[15] Wasilkowska A, Bartsch M, Messerschmidt U, Herzog R and Czyrska-Filemonowicz A 2003 *J. Mater. Proces. Technol.* **133** 218.
[16] Dubiel B, Wosik J, Penkalla H J and Czyrska-Filemonowicz A 2000 in Proc. of the Sterology and Image Analysis in Materials Science, edited by Fotobit Design, Kraków 135.
[17] Schröder J H and Arzt E 1988 *Praktische Metallography* **25** 264.
[18] Kusabiraki K, Takasawa Y and Ooka T 1995 *ISIJ International* **35** 542.
[19] Ducki K J 2010 Microstructural aspects of deformation, precipitation and strengthening processes in austenitic Fe–Ni superalloy, edited by Wydawnictwo Politechniki Śląskiej, Gliwice (in Polish).
[20] Rosenberg J M and Piehler H R 1971 *Metallurg. Trans.* **2** 257.

IOP Conf. Series: Materials Science and Engineering **35** (2012) 012008 doi:10.1088/1757-899X/35/1/012008

Mechanical and microstructural aspects of severe plastic deformation of austenitic steel

K Rodak[1], J Pawlicki [2] and M Tkocz[2]
[1] Department of Materials Science, Silesian University of Technology, Krasińskiego 8, Katowice 40-019, Poland
[2] Department of Materials Technology, Silesian University of Technology, Krasińskiego 8, Katowice 40-019, Poland

E-mail: kinga.rodak@polsl.pl

Abstract. The paper presents the effects of severe plastic deformation by multiple compression in the orthogonal directions on the microstructure and the mechanical properties of austenitic steel. Several deformation variants were conducted with different number of passes. FEM simulations were performed in order to evaluate the actual values of the effective strain in the examined, central parts of the compressed samples. The deformed microstructure was investigated by means of the scanning transmission electron microscopy (STEM) and the scanning electron microscopy (SEM) supported by the electron back scattered diffraction (EBSD). X-ray phase analysis was performed to evaluate the martensite volume fraction. The mechanical properties were determined by means of the digital image correlation method and hardness testing. It is shown that the applied forming technique leads to strong grain refinement in the austenitic steel. Moreover, deformation induces the martensitic $\gamma-\alpha'$ transformation. The microstructural changes cause an improvement in the strength properties. The material exhibits the ultimate tensile strength of 1560 MPa and the yield stress of 1500 MPa after reaching the effective strain of 10.

1. Introduction

Severe plastic deformation (SPD) techniques enable conventional macro-grained materials to be directly transformed into submicrometer and nanoscale materials. The final grain size and the mechanical properties depend on such factors as a material type, technique of deformation and process parameters. Austenitic stainless steels are extensively used in different industrial applications and it is important to control their microstructure and the mechanical properties. In the case of such materials, the mechanical properties can be increased by strain hardening, phase transformation of the strengthened austenite as well as precipitation hardening [1,2]. The martensitic transformation is related to the austenite instability at temperatures close or below the room temperature [3]. The microstructure transformation susceptibility is correlated to the stacking fault energy (SFE). The 304 austenitic steel exhibits rather low SFE value, of just 18 mJ/m^{-2}, and therefore deformation causes the development of cellular microstructure, twinning, and shear band formation [4,5]. Due to good formability, austenitic stainless steels can be easily cold formed. During cold forming process strain hardening phenomena occur.

The aim of the presented investigations was to determine the effect of deformation by multiple compression in the orthogonal directions on the microstructure and the mechanical properties of the

Cr-Ni stainless steel. The mentioned forming technique makes it possible to achieve high values of the total effective strain in a workpiece [6,7] and it is recognized as a method for obtaining ultrafine-grained materials [8].

2. Experimental

2.1. Cold forming

Multiple compression was conducted by means of the Gleeble MAXStrain system at the Institute for Ferrous Metallurgy, Gliwice. Before deformation, the material (commercially available 304 austenitic stainless steel) was solutionized at 1200 °C and the grain size of about 150 μm was obtained thanks to this treatment. Samples with the square cross section (10×10 mm) and the length of 27 mm, unrestrained lengthwise, were subjected to multiple compression with 90 deg rotation between subsequent passes. Several deformation variants were performed with different number of passes (4, 8, 16, 32 and 40). Anvils of 10 mm in length were used. The samples were deformed with the average strain rate of 0.5 s^{-1}. According to the experimental data acquired, the actual true strain values (calculated as the natural logarithm of the initial to final sample height ratio) in the subsequent passes were within the range of 0.15-0.3.

2.2. FEM simulations

Due to the presence of friction at the anvil-sample contact surface and the stiff material ends outside the deformation gap, distribution of the local effective strain is obtained in the sample after compression. In this study only the central regions of compressed samples are analysed. The highest strain accumulation occurs just in the centre of a sample and is greater than the true strain determined from the sample height reduction. In order to evaluate the actual total values of the effective strain in the samples centres after subsequent passes, numerical simulations corresponding to the experiment conditions were executed. The Forge2009 FEM software were applied.

2.3. Microstructure investigations

The JEOL 100B TEM microscope and the FEI INSPECT F SEM microscope with a field emission type gun and EBSD facility were employed for the microstructure investigations. In this study, TEM and SEM were conducted on samples extracted from the longitudinal plane section. Foils for TEM investigations were prepared using conventional techniques of electropolishing with the A8 electrolyte. For the SEM observations, specimens were polished using the ion thinning device (PECS) manufactured by Gatan Inc. Orientation maps were acquired with the step size of 50 nm.
X-ray phase analysis was performed in order to evaluate the effect severe plastic deformation on the martensitic transformation. The X-Pert Philips diffractometer equipped with the graphite monochromator were used. The diffraction profiles were obtained at the step size of 0.02°. The obtained diffraction patterns were analyzed in the X'PERT program.

2.4. Mechanical testing

Determination of the mechanical properties such as the yield strength (YS), the ultimate tensile strength (UTS) and the elongation at break (tEL) was performed on the MTS QTest/10 machine equipped with digital image correlation system (DIC) [9]. The DIC method is based on computational algorithms that track the grey value patterns in digital images of test surfaces, taken before and after an event that produces surface displacements. The precision of the method is of the order of 2/100 pixel while the minimum detectable displacement is of the order of 1/100 pixel. The 0.3 mm thick minisamples were used for tensile tests, with the gauge length of 1.0 mm and the total length of 2.2 mm. The hardness was determined according to the Vickers method and a load of 5 kgf was applied.

IOP Conf. Series: Materials Science and Engineering **35** (2012) 012008 doi:10.1088/1757-899X/35/1/012008

It should be noted that the mechanical properties were evaluated only to the central parts of compressed samples. Therefore, they are representative to the microstructures presented in the paper and correspond to the evaluated total values of the effective strain.

3. Results and discussion

Figure 1 illustrates shapes of the selected samples after compression and the effective strain distributions calculated by the applied FEM software. It is clearly seen that the distribution pattern remains similar for all the strain levels. The differences of the local, total effective strain values are significant. Strain at the sample centre is at least twice as much as in the undersurface regions. Therefore, it is very important to determine the strain values that correspond to the locations from which metallographic sections and minisamples for mechanical tests are taken. The effective strain values at the analysed region (the sample centre), for all conducted and aforementioned deformation variants, are collected in table 1.

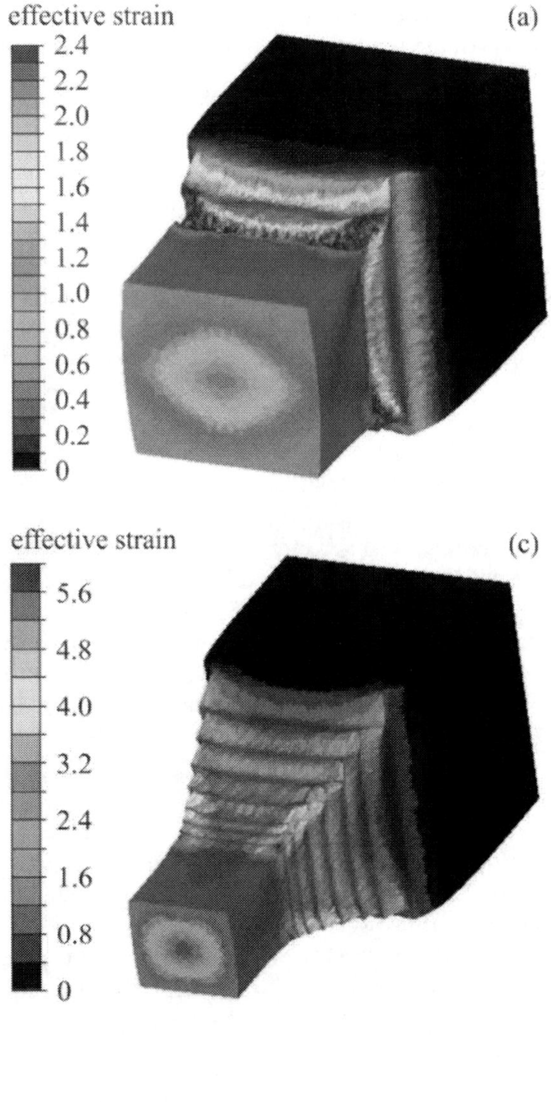

Figure 1. The effective strain distribution on halves of the compressed samples after 4 (a), 8 (b) and 16 (c) passes (the central cross-sections are seen at the front).

Table 1. Estimated total values of the effective strain at the centers of compressed samples.

Number of passes	Total effective strain
4	2.0
8	3.5
16	5.5
32	8.7
40	10.2

The effect of multiple compression on the microstructure of the investigated steel grade is presented in figure 2. STEM observations revealed the existence of a high density of dislocation cells (figure 2a) and numerous deformation twins in the central part of a sample after reaching the effective strain of 2. Significant differences in the microstructure were observed after obtaining the strain of 3.5. In these conditions the microstructure is composed of nanometer-thick lamellae of the martensite α' phase. The average transverse grain size is about 50 nm and the average longitude grain size of α' phase is a few hundred nanometers (figure 2b). Observations of the most deformed microstructure at the strain level of about 10 show the large areas of nano-sized, elongated grains (figure 2c). Extensively intersected boundaries of lamellae constructed a network pattern. The parallel plates characteristic for the α' martensite can be observed inside the small grains (figure 2d).

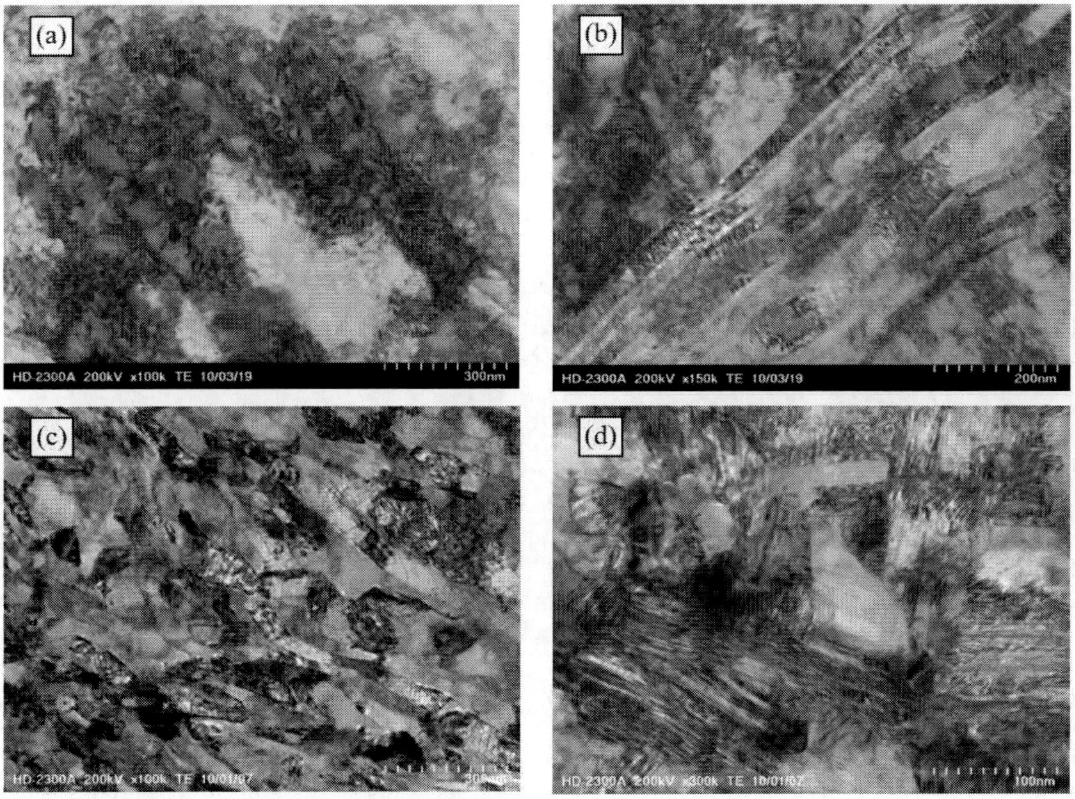

Figure 2. The 304 steel microstructure after reaching the effective strain of 2 (a), 3.5 (b) and 10.2 (c) and (d).

IOP Conf. Series: Materials Science and Engineering **35** (2012) 012008 doi:10.1088/1757-899X/35/1/012008

As presented in the EBSD orientation map (figure 3a), the most deformed sample centre (ε=10.2) exhibits wide variety of colors denoting very small, equiaxed and randomly oriented grains. This means that an increase of the effective strain results in generation of nano-sized grains. It should be noted that the very small equiaxed grains have a different crystallographic orientation (figure3a). It is evident that for the most deformed region of the sample (at its centre) the grain misorientation is high. The corresponding misorientation distribution is bimodal – the first distinct peak is observed for the low angles (that results from dislocation subboundaries) and another one for high angles (from 30° to 60°) (figure 3b). The average size of these grains varies from several tens of nanometers to submicrometers (figure 3c).

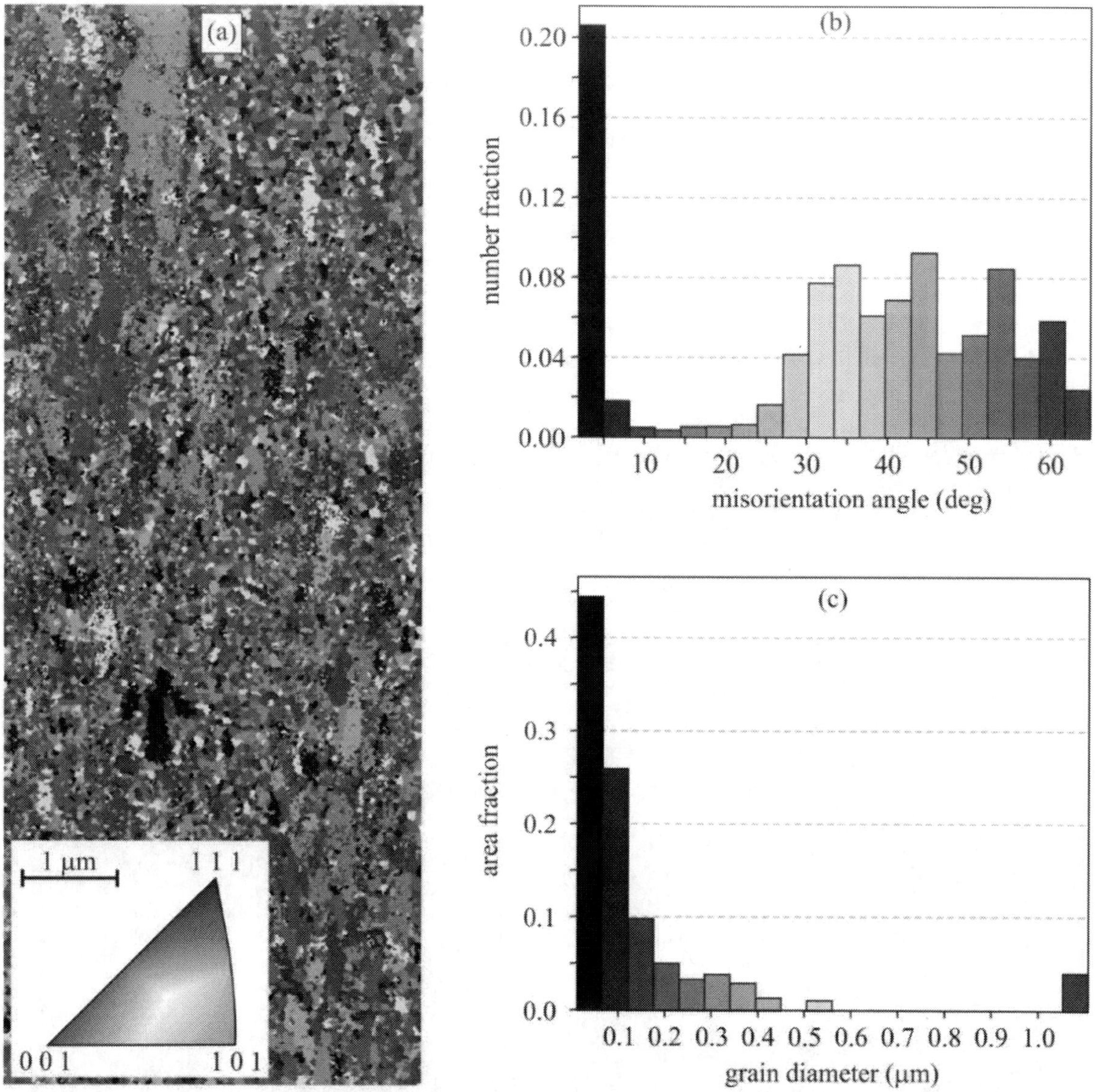

Figure 3. The EBSD results for the austenitic microstructure after multiple compression (ε = 10.2), a) The Inverse Pole Figure (IPF) map with colors corresponding to the crystal orientations shown in the legend, b) the misorientation distribution, c) the grain area fraction distribution.

The two-phase (α' and γ) microstructure of the examined Cr-Ni steel undergoes a significant change during deformation. The X-ray diffraction revealed that with the increase of effective strain within the range of 2 to 10, the amount of martensitic α' phase in steel rises from about 3% to about 55% (table 2). It was found that the deformation-induced phase transformation from austenite to martensite took place soon after deformation.

Table 2. The effect of severe plastic deformation on the change in martensite and austenite volume fractions in the investigated steel.

Total effective strain reached	Volume fraction (%)	
	Fe-α' (martensite)	Fe-γ (austenite)
2.0	3	97
5.5	22	78
8.7	42	58
10.2	55	45

The significant grain refinement and formation of the α' martensite are associated with a gradual increase of the mechanical properties in the investigated steel. The results of mechanical tests are presented in figure 4. It is clearly seen that multiple compression of the investigated alloy caused a significant strain hardening effect. After reaching the effective strain of 10, the ultimate tensile strength raised almost 3 times and the yield stress – almost 5 times. However, it is evident that the greatest strain hardening effect is obtained at the initial phase of deformation. The elongation at fracture decreases with the rise of effective strain, of course, however it still remains on a satisfactory level.

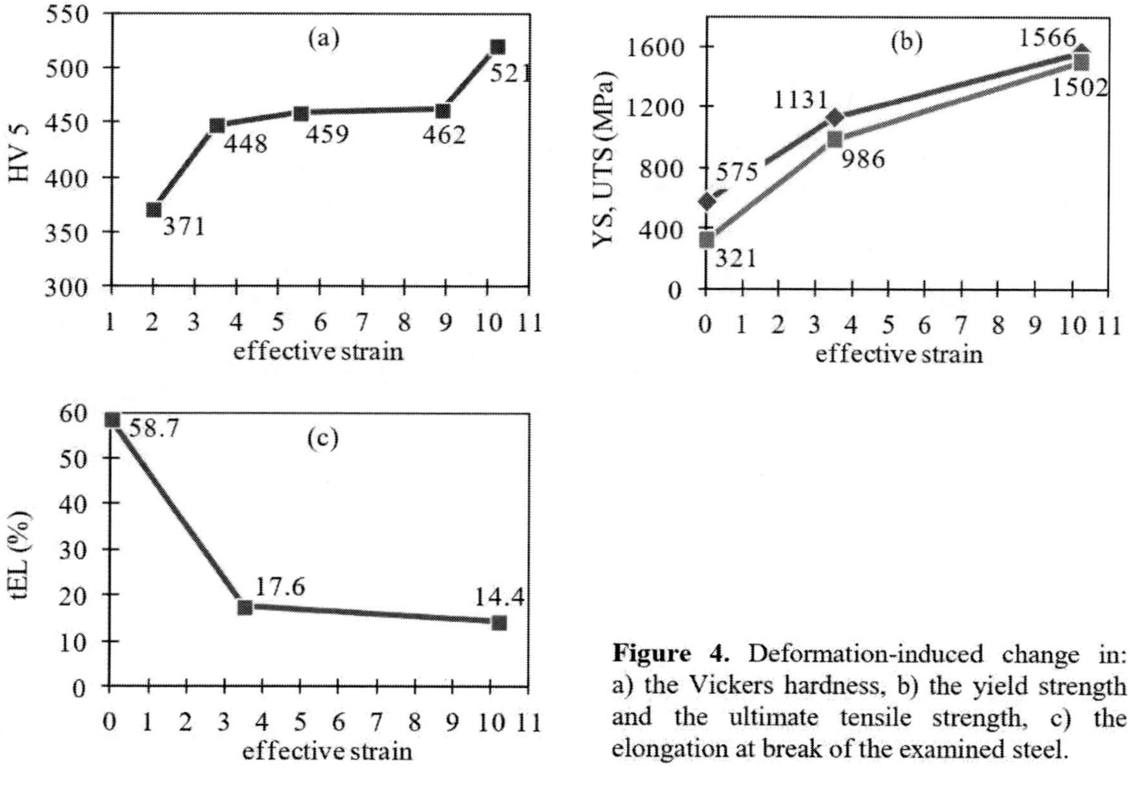

Figure 4. Deformation-induced change in: a) the Vickers hardness, b) the yield strength and the ultimate tensile strength, c) the elongation at break of the examined steel.

4. Conclusions

The results of presented investigations allow to draw the following conclusions:

- A character of the microstructure in the investigated steel strongly depends on a value of the effective strain. Severe plastic deformation conducted by multiple compression in the orthogonal directions causes a transition from a microstructure dominated by cells with low angle boundaries to an equiaxed nano-grained microstructure.
- The examined 304 Cr-Ni steel grade is the metastable one. Deformation induces the martensitic transformation within the whole range of applied strain.
- The amount of the α' phase also depends on the effective strain value. The martensite content of about 55% have been obtained in the sample region where the local, total effective strain reached about 10.
- The substantial grain refinement and the martensitic transformation cause the significant improvement in the strength properties of the examined steel and there is still some reserve of ductility.

References

[1] Solomon N and Solomon I 2010 *Revista de Metalurgia* **46** 121.
[2] Kurc A and Stokłosa Z 2010 *Arch. Mater. Sci. Eng.* **41** 85.
[3] Drahokoupil J, Hausild P, Davydov V and Pilvin P 2010 *Solid State Phenom.* **163** 295.
[4] Degtyarev M V 2007 *Acta Mater.* **55** 6039.
[5] Lewandowska M 2006 *Kształtowanie mikrostruktury i właściwości stopów aluminium metodą wyciskania hydrostatycznego* (Warszawa - Oficyna Wydawnicza Politechniki Warszawskiej).
[6] Petryk H, Stupkiewicz S 2007 *Mater. Sci. Eng. A* **444** 214.
[7] Kuziak R, Zalecki W, Węglarczyk S and Pietrzyk M 2005 *Solid State Phenom.* **101-102** 43.
[8] Petryk H 2008 *J. Mater. Process. Tech.* **204** 255.
[9] Chevalier L, Calloch S, Hild F and Marco Y 2001 *Eur. J. Mech. A-Solid* **20** 169.

IOP Conf. Series: Materials Science and Engineering **35** (2012) 012009 doi:10.1088/1757-899X/35/1/012009

Qualitative and quantitative description of microstructure of alloys from the Fe-Al system.

M Jabłońska[1], **M Mikuśkiewicz**[1] and **A Tomaszewska**[1]
[1] *Department of Materials Science, Silesian University of Technology,*
Krasińskiego 8, 40-019 Katowice, Poland

E-mail: magdalena.jablonska@polsl.pl, marta.mikuskiewicz@polsl.pl, agnieszka.tomaszewska@polsl.pl

Abstract. The paper presents the test results of qualitative and quantitative analysis of the structure of alloys from Fe-Al system after casting and heat treatment. The analysis were carried out for three alloys, with different content of Al at.%: 36, 38, 48 at %, which were produced by melting and gravity casting. A quantitative evaluation of the structure was made with use of "MET-ILO" application on the basis of images acquired from a light microscope. Moreover the influence of the chemical composition and results of quantitative description of microstructure on the hardness of alloys from the aluminium – iron system were analysed. The obtained research will be used for the development of mathematical models determining the influence of primary structure on the opportunities for plastic deformation of alloys. Structural examination was carried out using scanning electron microscopy (SEM) and scanning transmission electron microscopy (STEM). X-ray diffraction measurements were performed on this alloys.

1.Introduction

Selected alloys from the Fe-Al system are included into a group of materials on a matrix of intermetallic phases, and characteristic properties result from it and they constitute a resultant between properties of superalloys and ceramic materials. They are distinguished with individual features in a lattice structure, relating to pure components and precisely defined location of atoms in lattice nodes. The intermetallic phase is defined with the symbol AnBm, where a number of atoms in the component A is marked with a letter n, and a number of atoms in the component B is marked with a letter m. This combination enables to get a mixed character of interatomic bonds, where mostly covalent bonds are predominant and metallic bonds occur more sparsely. They form themselves as a result of link of metallic elements or a link of a metallic element with non-metallic element at strictly defined concentration of components, i. e. in a form of secondary solid solutions and they are distinguished with ordered internal structure. This phenomenon results from energy of occurring atom bonds, where energy between atoms of different elements is bigger than between atoms of the same elements [1].
The iron aluminides FeAl have been among the most widely studied intermetallics because their low cost, low density, good wear resistance, ease of fabrication and resistance to oxidation and corrosion create wide prospects for their industrial applications, for components of machines working at high temperatures and in corrosive environments [2-4]. This paper describes results of qualitative and quantitative analysis of the structure.

2.Experimental

The alloys comprised respectively 36%, 38% and 48% aluminium and they comprised alloy additives of modifying activity in a form of molybdenum, zirconium, boron and carbon as well. Material for the research consisted of bars cast from an alloy based on an FeAl intermetallic phase of the chemical composition shown in Table 1.

Table 1. Chemical composition of alloys from the Al-Fe system, applied for the tests.

Composition (%at.)	Al	Fe	Mo	Zr	C	B
Fe36Al	36.00	63.64	0.2	0.05	0.10	0.01
Fe38Al	38.00	61.64	0.2	0.05	0.10	0.01
Fe48Al	48.00	51.64	0.2	0.05	0.10	0.01

The samples were prepared by melting in spinel Al2O3×MgO crucibles in an induction furnace at vacuum. The following contents were used for smelting: ARMCO iron, aluminium 99.98 at.% minimum, amorphous boron and a technically-pure molybdenum powder compact. The got casts underwent homogenizing annealing in an induction furnace type CARBOLITE Sheffield England. Annealing in temperature 1000°C for 24 hours and cooling in air was applied for all alloys. Observation of the structure was carried out in use the optical stereoscope microscope SZX9 Olympus and light microscope OLYMPUS GX51. XRD measurements were performed in the 10–90° 2θ range with Cu K_ radiation and an incident beam monochromator (X'pert, Philips-Panalytical). A Retvield refinement was applied to the XRD results from a continuous step scan (0.0080° 2θ step size and 957.6 s scan time) to calculate the unit cell parameters. Microstructure was investigated in a scanning electron microscope (SEM, HITACHI S-3400N) equipped with an energy-disperse spectrometer (EDS, Thermo NORAN+System SIX). Alloys for TEM study were examined in a Scanning Transmission Microscope (Hitachi HD-2300A) with 200kV acceleration voltage.
A quantitative analysis was made by a surface method, using the computer program Met-Ilo v.12.1. Methodological problems in the analysis of the structural images obtained, consisting in the occurrence of artefacts, were solved by application of appropriate image transformations. During the quantitative analysis, twin boundaries were partly removed in an automatic way. A number of stereological parameters were determined, characteristic of single-phase structural materials, of which the following are presented in this publication [5,6]:
– mean area of grain plane section $\bar{A}$, μm^2
– index variability of the grain plane section area v(A), %
– length of grain boundaries, μm
– amount of grain on the unit area Na, mm^{-2}.

3. Results and discussion

The structure of the FeAl alloy investigated after casting and annealing was coarse-grained. In a case of the alloy Fe36Al with micro-additives, the analysis of structure in a state after casting, revealed occurrence of three crystalline zone (Fig.1a): a narrow zone of frozen grains, occurring only nearby surface of heat exchange (liquid alloy – mould), zone of column crystallites and internal zone of equiaxed grains. Moreover we observed influence of crystallization conditions one the structure. Bottom section of ingots contains wider zone of column crystallites than top section. It was observed for alloys with 36%, 38% and 48% at. aluminium. The situation was almost the same for alloy Fe38Al with micro-additives (Fig.1b). Together with increase in content of aluminium up to 48, homogenization of structure was noticed (Fig.1c). It could be caused by worse casting properties for alloy wit higher Al content.

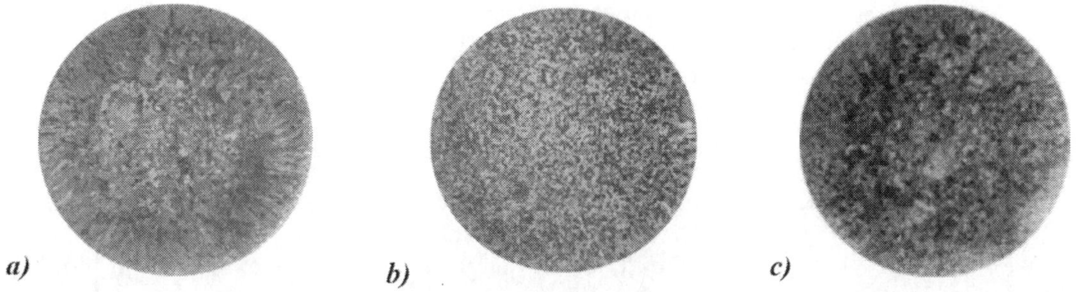

Figure 1. The structure of the FeAl alloy investigated after casting a) alloy Fe36Al with micro-additives in a state after having been cast. Three crystalline zones (from outsider: fine frozen crystallites, column crystallites and equiaxed crystallites), b) alloy Fe-38Al with micro-additives in a state after having been cast, c) alloy Fe48Al with micro-additives in a state after having been cast.

The analysis of microstructure LM, in a state after casting for alloys with 36% at. (Fig. 2a), 38% at. (Fig. 2b), 48% at. (Fig. 2c) aluminium, revealed the coarse-grained structure. The grains are characterized by changeable shape, size and non-continuity on boundaries. By supplementing a word description in qualitative methods with a set of quantitative parameters allows characterising the structure of alloy in an objective and precise manner.

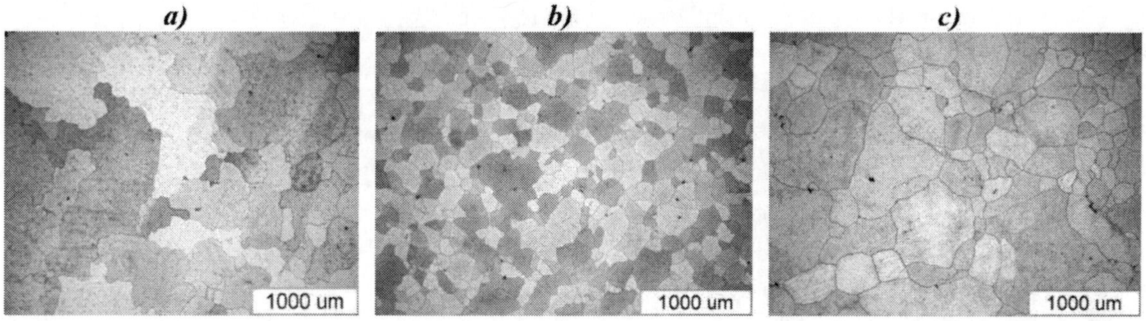

Figure 2. Microstructure LM of the alloy a) 36 at.% Al, b) 38 at.% Al, c) 48 at.% Al with micro-additives after annealing at 1000°C for 24h and cooling on air.

Table 2. The results of stereological parameters determined Fe-Al alloys structure quantitative evaluation.

	Fe36Al	**Fe38Al**	**Fe48Al**
Mean area of grain plane section $\bar{A}$, μm^2	143137	16414	36471
Index variability of the grain plane section area ν(A), %	146.7	109.9	152.5
Length of grain boundaries, μm	94640	258647	169187
Amount of grain on the unit area Na, mm^{-2}	6.90	58	26.6

The analysis of the structure of the tested samples showed that all parameters has a diversified size depending on Al content. An analysis of the results of grain quantitative evaluation presented in Table 2. The alloy with 36 at%. Al is characterized by the largest size of the grains, whose mean plane section area of grains equals $\bar{A}$ = 143137 μm^2 with a variability index ν($\bar{A}$) = 146.7 % while the alloy with 38 at%. Al is characterized by the smallest size of the grains, whose mean plane section area of grains equals Ā=16414 μm^2 with a variability index v(Ā)=109.9 %. Similar tendencies refer to the amount of grain on the unit area (respectively, Na for Fe36Al = 6.90 mm^{-2}, Fe38Al = 58 mm^{-2}) and length of grain boundaries (94640 μm to Fe36Al and 258647 μm to Fe38Al). Along with an increase Al content to 48%, mean area of grain plane section increase (Ā=36471 μm^2) and the other parameters decrease (length of grain boundaries=169187 μm and Na=26.6 mm^{-2}).

Studies in the scanning electron microscope (Fig. 3 a,b,c.) disclosed a lot of small precipitations of micro-additives inside the grains and on the grains boundaries for the all investigated of alloys (Fig. 4). It was established that for all alloys the best results of microstructures were subjected to chemical etching in 10 ml HNO_3, 10 ml HF, 15 ml H_2O solutions and reorder chemical etching in 11 ml HCl, 22 ml HNO_3 and H_2O.

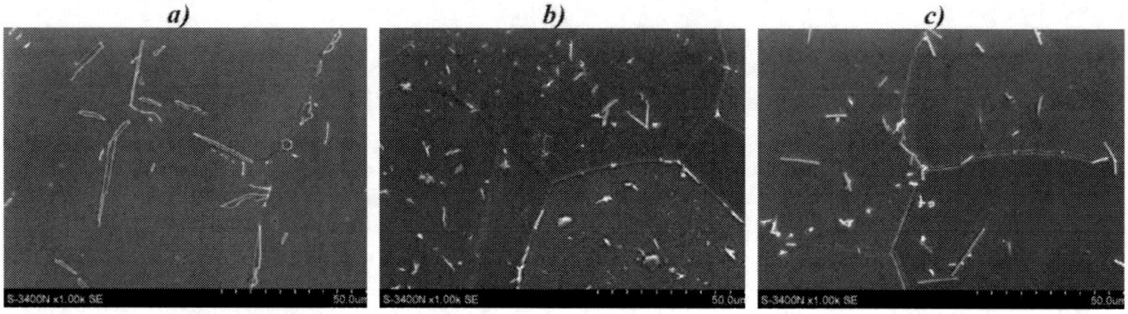

Figure 3. Microstructure SEM of the alloy a) 36 at.% Al, b) 38 at.% Al, c) 48 at.% Al investigated after annealing at 1000°C for 24h and cooling on air.

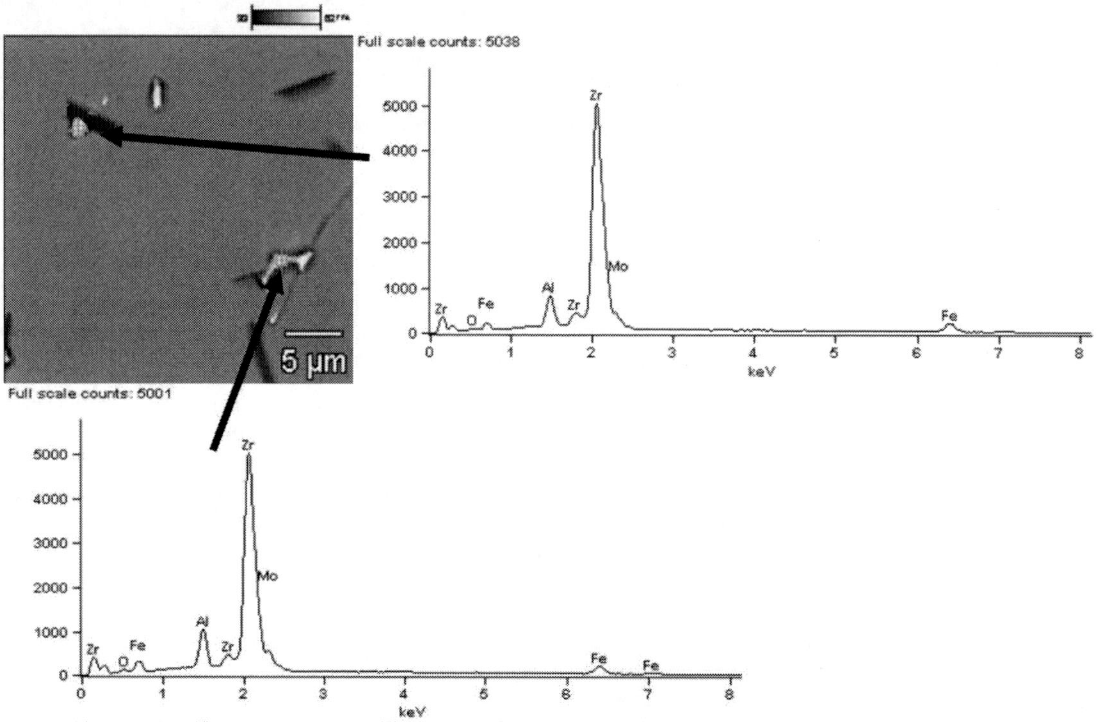

Figure 4. Microstructure (SEM) and EDS in the indicated area of alloy with 48 at.% Al.

TEM investigations were performed in order to evaluate the grain size and further study ordering state of the alloys. Studies in the transmission electron microscope confirmed that an increase of at. % Al content to an increase of density of dislocation cells (see Figure 5).

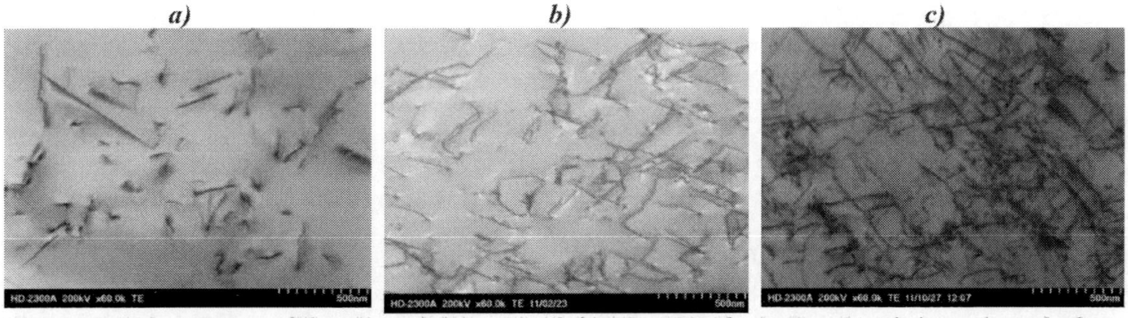

Figure 5. Substructure of the alloy a) 36 at.% Al, b) 38 at.% Al, c) 48 at.% Al investigated after annealing at 1000°C for 24h and cooling on air.

No significant changes of the phase composition were observed for alloys with different aluminum content. The X-ray analysis reveals that all samples contained the following main phase: Fe-Al, Fig. 6 a, b, c (ICDD PDF 03-065-6132). Results of the unit cell obtain by Rietveld method for alloys with 36, 38 and 48 at.% Al content are shown in Table 3. In this case was observed the insignificant increase of unit cell parameters (Fe36Al). For the rest of samples confirmed more deformed structure – all the examined parameters of the unit cell.

Table 3. Parameters of the unit cell [Å] FeAl after calculation by the Rietvield method.

	ICDD	Fe36Al	Fe38Al	Fe48Al
			Rietveld	
a_0 [Å]	2.9090	2.9012	2.9158	2.9125
b_0 [Å]	2.9090	2.9012	2.9158	2.9125
c_0 [Å]	2.9090	2.9012	2.9158	2.9125

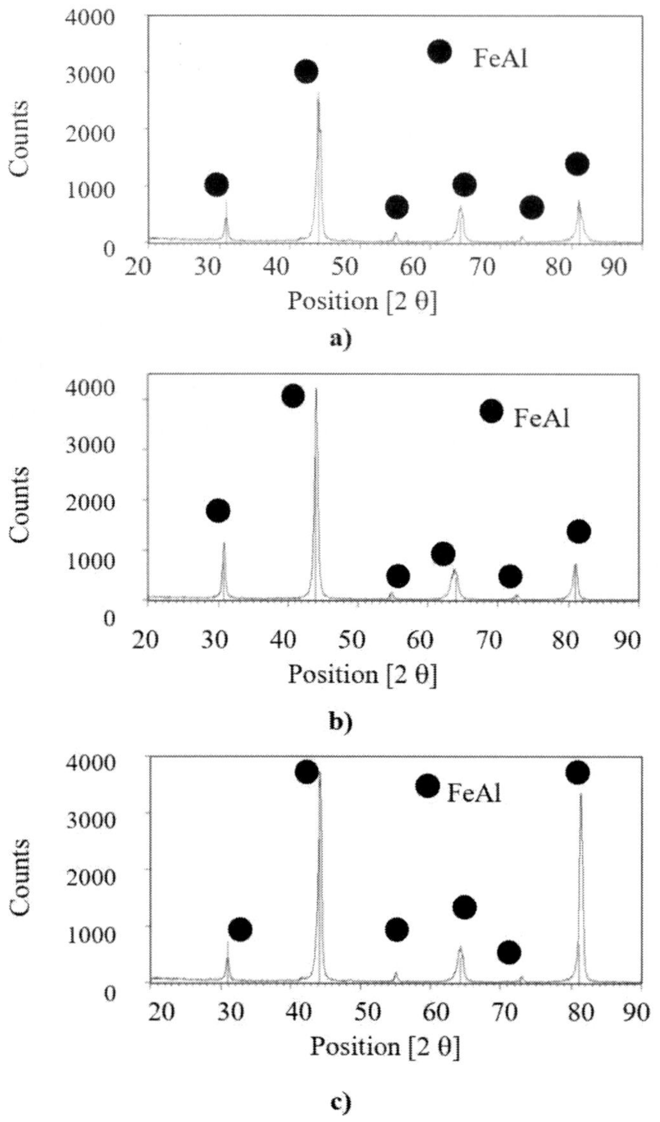

Figure 6. X-ray diffraction spectra of alloys after annealing at 1000°C for 24h and cooling on air.

4. Summary

In this work, qualitative analysis of the structure of intermetallic compound samples Fe-Al were determined by was investigated in a scanning electron microscope (SEM, HITACHI S-3400N) equipped with an energy-disperse spectrometer (EDS, Thermo NORAN+System SIX) and scanning transmission microscope (Hitachi HD-2300A). Quantitative analysis of the structure determined by "MET-ILO" application on the basis of images acquired from a light microscope.

In summary, the following conclusions can be given:

- FeAl phase of B2 type structure appeared to be main phase of the studies samples,
- The metallographic tests revealed occurrence of coarse-grained structure of different morphology in all tested alloys. In the studied samples presence of different size of the grain depending on Al content was proved. The alloy with 36 at%. Al is characterized by the largest size of the grains ($\bar{A}$ = 143137 μm^2) while the alloy with 38 at%. Al is characterized by the smallest ($\bar{A}$=16414 μm^2). The same tendencies refer to the amount of grain on the unit area

IOP Conf. Series: Materials Science and Engineering **35** (2012) 012009 doi:10.1088/1757-899X/35/1/012009

and length of grain boundaries. Along with an increase Al content to 48%, mean area of grain plane section increase contrary to length of grain boundaries and Na,

- In the all investigated of alloys revealed a lot of small precipitations of micro-additives inside the grains and on the grains boundaries,
- The lattice parameter slightly increases with the increase of aluminum content. This fact is in agreement with structural heterogeneity with different recrystallisation degree and different grain size.

Performance of next tests on alloys of higher aluminium content is expected.

Acknowledgement
Financial support of Structural Funds in the Operational Programme – Innovative Economy (IE OP) financed from the European Regional Development Fund - Project No POIG.01.01.02-00-015/09 is gratefully acknowledged.

References
[1] Niewielski G and Jabłońska M 2007 *Mater. Eng.* **2** 43.
[2] Stoloff NS 1998 *Mater. Sci. Eng.* **A 258** 1.
[3] Chakraborty P, Sharma IG, Suri AK and Bose DK 2001 *J. Mater. Process. Technol.* **115** 413.
[4] Sastry DH, Prasad YVRK and Deevi SC 2001 *Mater. Sci. Eng.* **229** (1) 63.
[5] Kurzydłowski KJ and Ralph B 1995 *CRC Press Baton Rouge, USA.*
[6] Cwajna J, Maliński M and Szala J 1993 *Mater. Sci. Eng.* **3** 79.

IOP Conf. Series: Materials Science and Engineering **35** (2012) 012010 doi:10.1088/1757-899X/35/1/012010

Degradation of Si-Al aluminide coating after service of turbine blades made of ZhS6K superalloy

B Chmiela[1], M Kianicová[2], M Sozańska[1] and L Swadźba[1]
[1] Department of Materials Science and Metallurgy, Silesian University of Technology, 40-019, Katowice, Poland
[2] Faculty of Special Technology, Alexander Dubcek University in Trenčín, Študentská 1, 911 50, Trenčín, Slovakia

E-mail: bartosz.chmiela@polsl.pl

Abstract. Aero engine turbine blades made of nickel-based superalloys are characterized by very good mechanical properties, but their hot corrosion resistance is insufficient. Therefore, various protective coatings must be applied. These coatings are typically made of diffusive aluminide coatings based on theβ-NiAl intermetallic phase. Although the oxidation resistance and hot corrosion resistance of these coatings are very good, their thermal resistance is relatively poor. As a result, turbine blades with aluminide coatings are prone to degradation in case of overheating. In this paper we study the degradation of the Si-Al aluminide coating on turbine blades made of ZhS6K superalloy during overheating in the DV2 jet engine.

1. Introduction

Aero engines turbine blades are flight safety parts and are produced by the investment casting method using ceramic molds [1,2]. During service, creep, thermo-mechanical fatigue and hot corrosion processes occur [3]. Hot corrosion is caused by the mixture of many compounds forming low-melting-point mixtures (NaCl, Na_2SO_4, SO_2, SO_3), which leads to failure of the blade surface [4–6]. Although the mechanical properties of turbine blades are very high (as a result of proper chemical and phase composition), the hot corrosion resistance is insufficient. Therefore, to improve hot corrosion resistance and oxidation resistance, turbine blades are coated with protective coatings [6,7]. Commonly used are diffusive aluminide coatings based on the β-NiAl intermetallic compound with a zonal structure. Aluminide coatings are characterized by excellent oxidation resistance and hot corrosion resistance, but they suffer from poor thermal resistance. Analogous to this $TiAl_3$ coating deposited onto Ti alloy exhibits a significant influence on a reduction of the oxidative abilities of Ti-based alloys due to properties of the coating (chemical stability, thickness, negligible porosity, and strength of substrate-coating chemical bond) [8]. Therefore, for turbine blades with aluminide coatings, overheating may be very dangerous [7]. In this paper, we investigate the degradation of turbine blades made of ZhS6K nickel-base superalloy with a Si-Al aluminide coating after overheating as a result of compressor stall in the DV2 jet engine [7,9].

2. Material and experimental procedure

The investigated hollow turbine blades were a precision casting made of ZhS6K nickel-base superalloy. The chemical composition of this alloy is given in table 1.

Table 1. Chemical composition of ZhS6K superalloy (wt. %).

Ni	Cr	Co	Mo	W	Al	Ti	Fe	Mn	Si	C	B	Zr
balance	10.5	5.2	4.1	4.7	5.5	2.8	1.8	0.4	0.4	0.15	0.035	0.04

After solidification, a solution heat treatment was performed. Then the protective aluminide coating (Si-Al) was deposited on the blade surface (table 2) [7].

Table 2. Heat treatment and chemical-thermal treatment of ZhS6K superalloy.

Type of treatment	Procedure
homogenization	vacuum, 1225°C, 4 h, cooling by argon stream at an overpressure of 1.5–2 bar
Annealing to remove residual stress after mechanical treatment	950 °C, 2 h dwell at temperature in dry argon; cooling in air
Si-Al layer deposition	stage 1: spraying a layer (350 mL coloxyline solution, 112 g Al powder, 112 g Si powder) stage 2: diffusion annealing at 1000 °C 3 h dwell at temperature; slow cooling in a retort.

The DV2 jet engine is characterized by a service life of 1500 h. The nominal turbine inlet temperature is in the range of 800 to 1050 °C, but servicing at high temperature is acceptable only for a short time. The maximum acceptable exhaust-gas temperature is 760 °C. For the engine investigated, the compressor stall happened (unstable working) after 167.6 h of service. As a result, the high-pressure turbine overheated for 12 s and the highest exhaust-gas temperature was 977 °C (figure 1) [7].

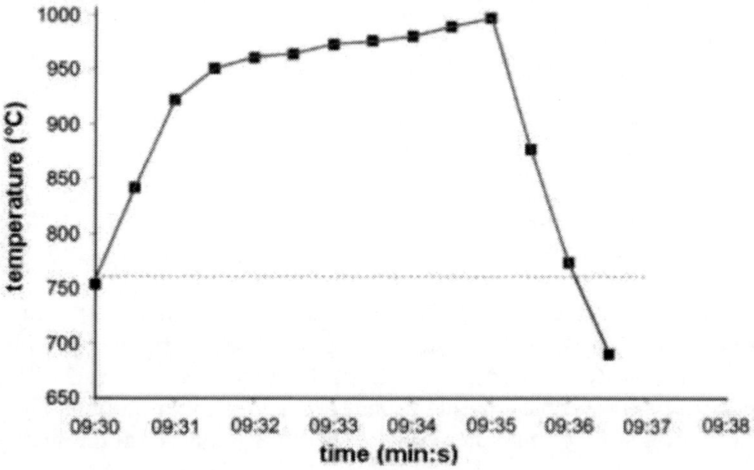

Figure 1. Critical temperature excess of DV2 engine.

The difference between maximum and acceptable exhaust gas temperature and overheating time were used to evaluate the blade degradation using the parameter D [11]:

$$D = (t_2 - t_0)(T_{max} - T_C)^3 \int_{t_0}^{t_2} [T(t) - T_C] dt \qquad (1)$$

where: $t \in \langle t_0, t_2 \rangle$ is the overheating duration defined by $T > T_C$ and T_{max} is the maximal overheating temperature (°C), $T(t)$ is actual engine temperature in the time t (s).

On the basis of the parameter *D*, it is possible to predict of the degree of blade degradation [7] (table 3).

Table 3. Degree of blade degradation as a function of *D*.

D (°C^4·s^2)	engine status	performance capability
$1.0\cdot10^5 \div 1.5\cdot10^9$	full performance	service
$1.5\cdot10^9 \div 1.0\cdot10^{10}$	degradation of coating of high-pressure turbine blade	repair
$1.0\cdot10^{10} \div 1.0\cdot10^{12}$	degradation of substrate of high-pressure turbine blade	repair
$1.0\cdot10^{12} \div 1.0\cdot10^{14}$	degradation of substrate of low-pressure turbine blade	repair
$1.0\cdot10^{14} \div 2.2\cdot10^{14}$	degradation of basic material of discs	repair
$> 2.2\cdot10^{14}$	blade melting	out of service

After service, the DV2 engine was disassembled to investigate the surface and cross-section of the high-pressure turbine blade. The microstructure, chemical composition, and phase composition were evaluated using scanning electron microscopy (SEM, Hitachi S-4200), a Hitachi S-3400N equipped with an energy-dispersive spectrometer (EDS, Thermo NORAN (System Six)), and electron backscatter diffraction (EBSD, INCA HKL Nordlys II (Channel 5)). The turbine blade was cut perpendicular to the main axis at 1/3 of the airfoil length from the tip, because in this area the interaction with hot combustion gases was the strongest [7]. Next, the specimen was mounted, ground on abrasive paper, and polished using diamond and alumina suspensions. The cross section was investigated in the four important areas: the leading edge, the trailing edge, the pressure side, and the suction side (figure 2). For metallographic examination, the specimen was etched in a solution of 100 cm^3 HCl, 100 cm^3 HNO_3, 100 cm^3 H_2O, and 3 g MoO_3.

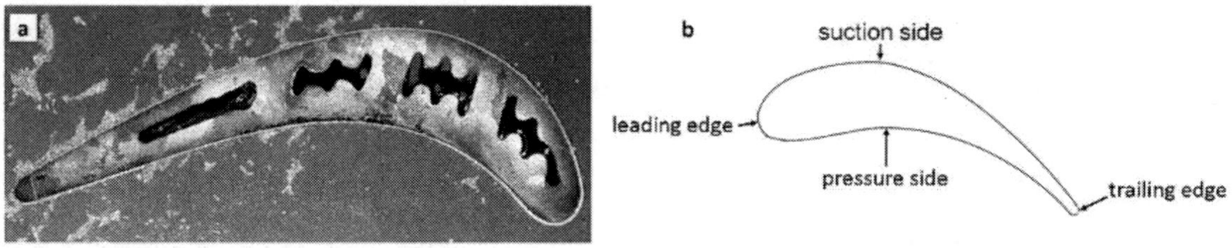

Figure 2. Cross section of blade (a) and areas investigated on cross section of blade (b).

3. Results and discussion

Investigations of the blade surface reveal the appearance of the aluminide coating on the airfoil. Macroscopic observations show that the coating damage is stronger on the suction side (convex side of the airfoil) than on the pressure side (concave side of the airfoil, see figure 3). Visible failures seem not to be very large, but microscopic investigations reveal the real condition. The main cause of the observed damage may be a stronger interaction of hot combustion gases at the suction side.

IOP Conf. Series: Materials Science and Engineering **35** (2012) 012010 doi:10.1088/1757-899X/35/1/012010

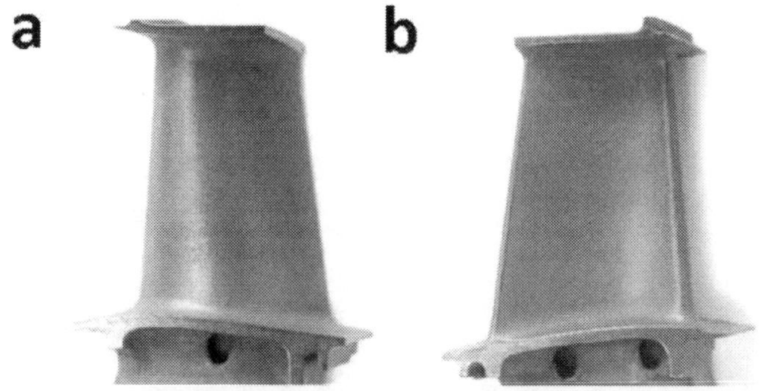

Figure 3. Surface of aluminide coating on the blade airfoil: (a) suction side, (b) pressure side.

SEM investigation revealed a scale on the airfoil surface, cracks and spalling areas (figure 4).

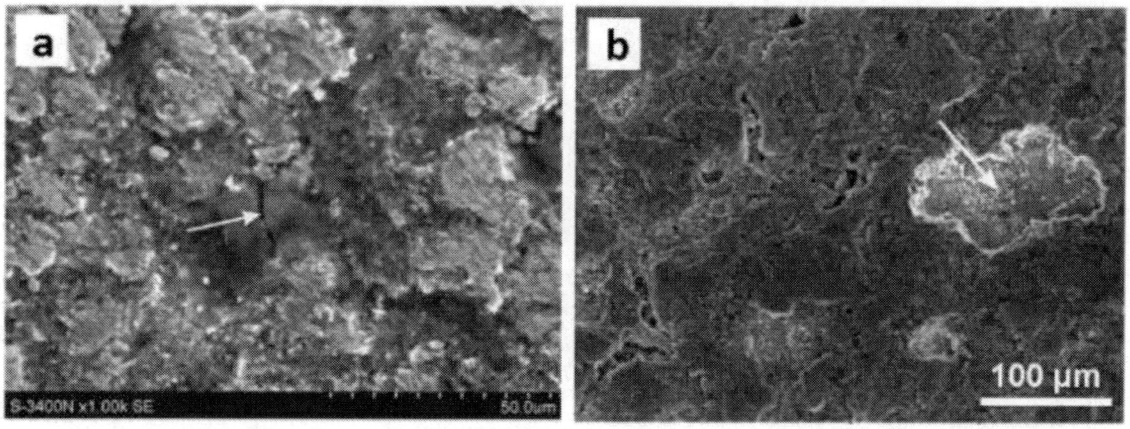

Figure 4. The crack on the blade surface (a) and spalling of the part of blade coating (b).

EDS analysis confirmed the presence of the oxide scale on the suction side and on the pressure side (figure 5).

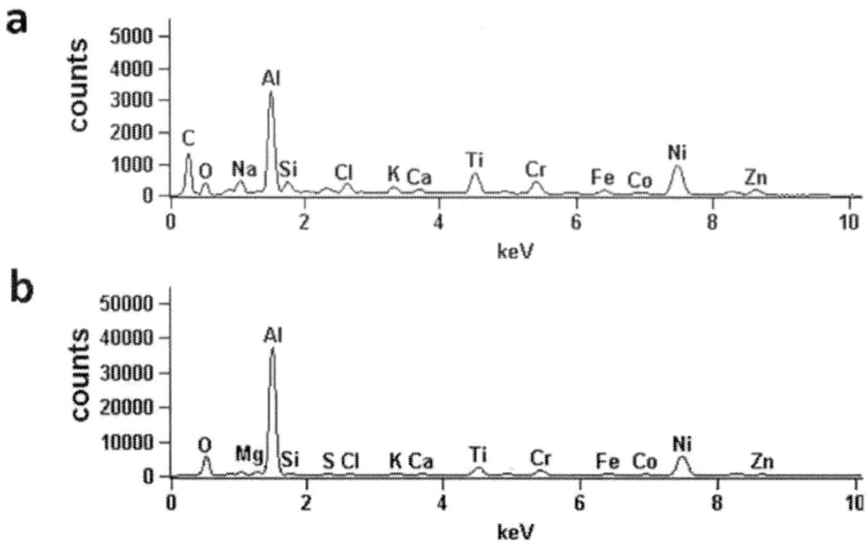

Figure 5. EDS spectra of oxide scale on the airfoil surface: (a) suction side, (b) pressure side.

Qualitative results of EDS analyses clearly suggest the presence of aluminum oxide. Moreover, the other elements identified in the scale (sulphur, chlorine, sodium, potassium, calcium) indicate an intensive hot corrosion process and degradation of the coating.

On the basis of temperature (T_{max}) and overheating time, the parameter D was calculated to be $3.2 \cdot 10^{11}$ °C$^4 \cdot$s^2, which indicates (Table 1) a high possibility of degradation of the coating and the substrate (ZhS6K). Investigations into the cross section of the airfoil reveal the real degree of material degradation (figure 6 and figure 7). Sudden cooling of blade surface from high temperature together with cooling of substrate by means of internal blade cooling channels lead to the generation of tensile stresses in the coating (from reason of misfit of coefficient of thermal expansion coating and substrate) and then cracks perpendicular to the coating - substrate interface.

Along with cooling of blade surface, condensation of exhaust gases appears and as a result the corrosion enhances. These sites serve as starters for creation and propagation of cracks not only perpendicular to the interface, but also parallel with interface (figure 7). The final consequence of this mechanism is spalling of the scale (figure 5).

Turbine blades of DV2 engine did not correspond to the state after overheating from the view of thickness of coating (fig. 7). Overheating was caused by incomplete combustion in the chamber, i.e. impropriate proportion of fuel and air (for example in the consequence of surging) what causes burning of mixture and high temperature in the low-pressure turbine in the vicinity of thermocouples.

This case can be easily revealed by monitoring of parametric engine data as revolutions and relating fuel flow.

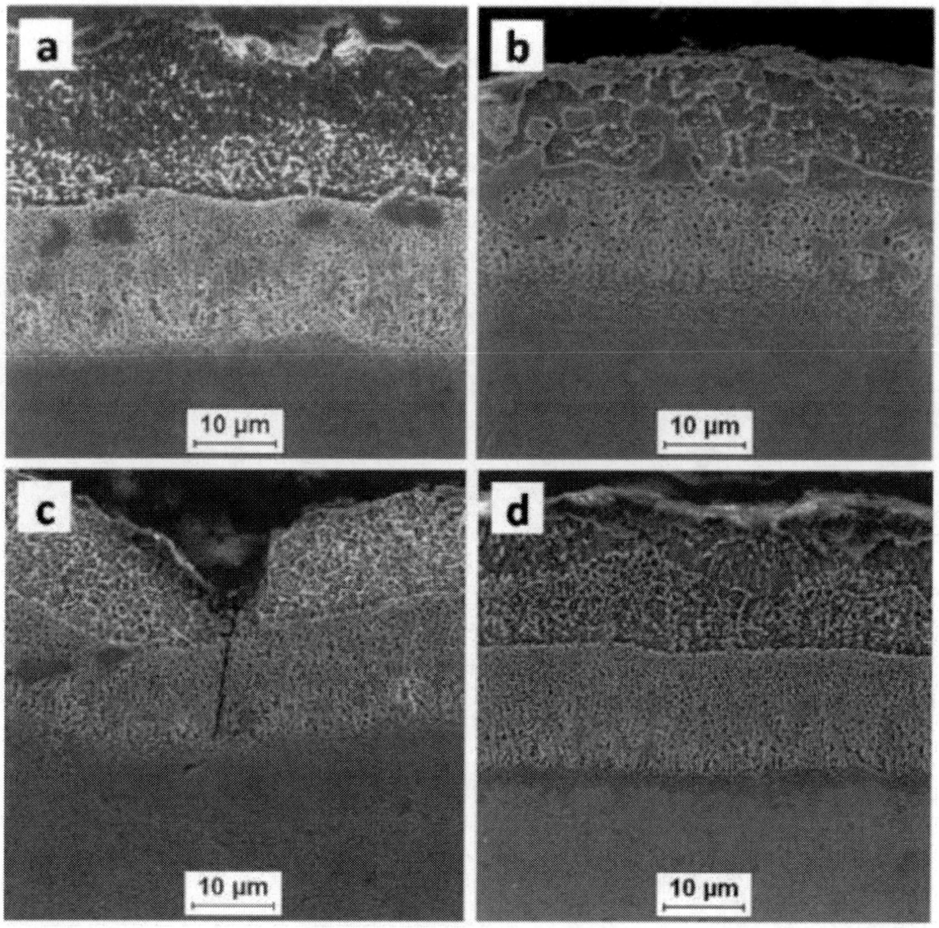

Figure 6. SEM images of the cross section of the blade airfoil: (a) leading edge, (b) trailing edge, (c) suction side, (d) pressure side.

The most damaged area was the suction side. SEM images reveal hot corrosion pits, which expose the surface of the ZhS6K substrate to the high temperature and combustion gases. Moreover, some cracks transpiercing the entire aluminide coating are visible. These cracks allow combustion gases to damage the substrate, too. On the pressure side of the airfoil only several cracks perpendicular to the surface were found. The leading edge and the trailing edge display slight damage.

SEM investigations revealed many needle-shaped precipitates just under the aluminide coating. EDS analysis confirms a high concentration of tungsten, molybdenum, and chromium (figure 8), which is characteristic for topologically close packed (TCP) phases. Precipitation of TCP phases in superalloys is typical for long thermal exposure or overheating. TCP phases are usually brittle and decrease the mechanical properties of turbine blades [1,2].

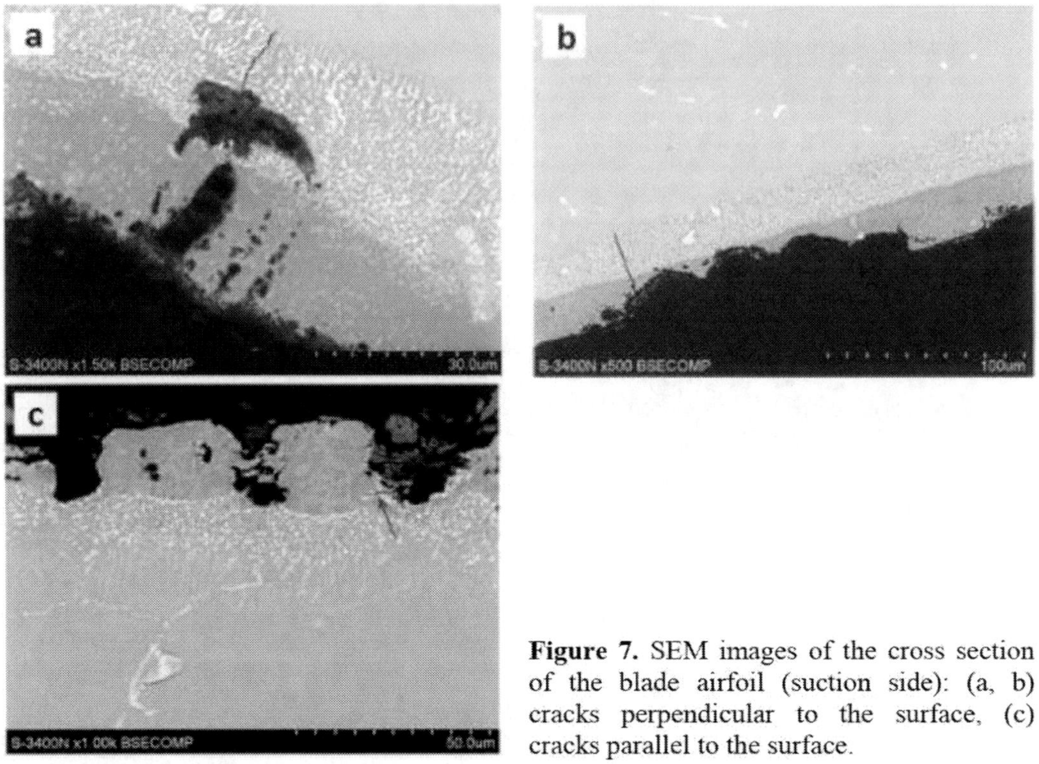

Figure 7. SEM images of the cross section of the blade airfoil (suction side): (a, b) cracks perpendicular to the surface, (c) cracks parallel to the surface.

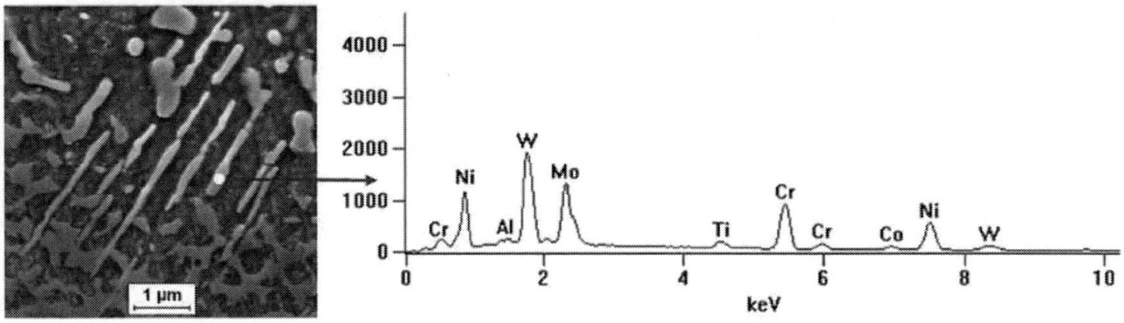

Figure 8. EDS analysis of needle-shaped precipitates under aluminide coating.

Because of small size of these precipitates, we applied the EBSD technique to identify them [10]. EBSD analysis confirms that needle-shaped precipitates are mainly σ phase (figure 9a).

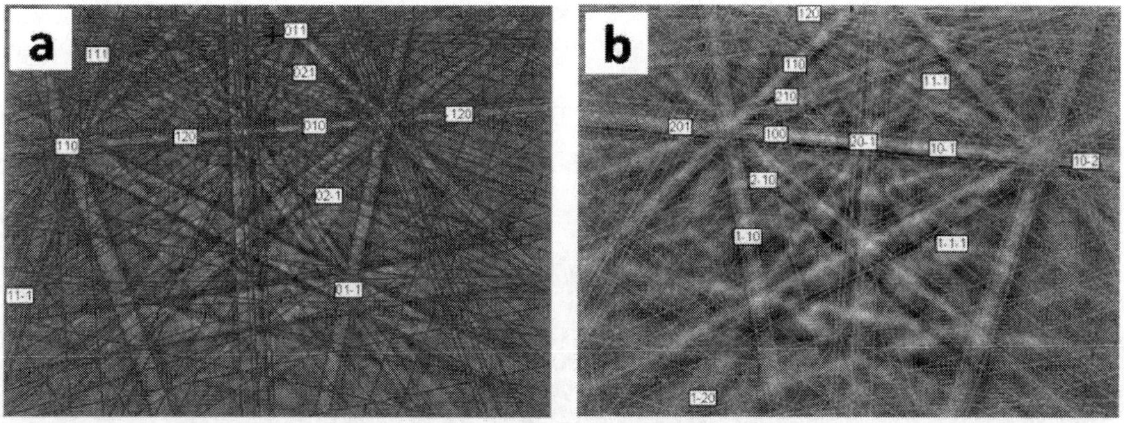

Figure 9. Kikuchi patterns of needle-shaped precipitates under the aluminide coating: (a) σ phase, (b) probably P phase.

Some precipitates are tentatively identified as P phase (figure 9b). This case clearly reveals that EBSD analysis may be ambiguous. The main reasons are (i) the very similar unit cell parameters of σ and P phases and (ii) the small size of precipitates, near the limit of spatial resolution for the EBSD technique [10].

5. Conclusions

We investigated degraded high-pressure turbine blades and found that blades with diffusive aluminide coating are very sensitive to overheating caused by unstable working of the engine. Although oxidation resistance of a Si-Al coating is quite good (due to high aluminum content), the thermal resistance is generally poor. Overheating in the short time (12 s) in connection with corrosive combustion gases caused degradation of the coating by forming pits and cracks. Some cracks transpierce the coating and expose the substrate, which may be a reason for ZhS6K superalloy damage. Moreover, many precipitates of TCP phases appear just under the coating, which confirms that the temperature was too high in this area. All these factors are very dangerous and clearly reveal that further service of the engine is very risky, although its service life is not exceeded.

The microstructure of the blade does not correspond to the expected results which should lead to a high degree of degradation due to the high values of overheating parameters. Obviously, the performance capability class (degraded base material of ZhS6K blades and replacement of high-pressure turbine blades) was incorrectly determined by the manufacturer. On the basis of the criteria determined, the structure analyses of blades have confirmed the satisfactory state of blades.

This fact led to a repeated examination of the event and all the overheating parameters. The re-examination of local analysis data revealed that the cause of increased critical temperature of outlet gases was deceleration, i.e. the flame extinction in the combustion chamber, which led to a drop in the revolutions of high-pressure rotor below the nominal value of 53% (idling).

The reaction of the pilot (trying to increase these revolutions) led to shifting the engine control lever towards higher degrees, which increased the flow of fuel into the no-longer burning combustion chamber. Thus a state was reached in the combustion chamber when fuel could not be ignited due to the low air excess coefficient α.

The ignitability of the mixture rich in fuel was renewed where it got mixed with air that cools the turbine (beyond the first or second stage of low-pressure turbine) depending on what volume of air the mixture needed to be in the combustion area of the quenching characteristic. The combustion outside the combustion chamber led to a disinformation about the temperature of the operating blades of high-pressure turbine and it explains the satisfactory microstructure condition of the creep-resistant coating.

Acknowledgments
Financial support of Structural Funds in the Operational Programme - Innovative Economy (IE OP) financed from the European Regional Development Fund - Project "Modern material technologies in aerospace industry", No. POIG.01.01.02-00-015/08-00 is gratefully acknowledged. This work was also supported by the Ministry of Industry and Trade of the Czech Republic in the frame of the Project FR-TI1/237 099.

References

[1] Reed R C 2006 *The Superalloys Fundamentals and Applications* (Cambridge: University Press)
[2] Durand-Charre M 1997 *The Microstructure of Superalloys* (Amsterdam: Overseas Publishers Association)
[3] Boyce M P 2002 *Gas Turbine Engineering Handbook* (Houston: Gulf Professional Publishing)
[4] Eliaz N, Shemesh G and Latanision R M 2002 *Eng. Fail. Anal.* **9** 31
[5] Deb D, Iyer S R and Radhakrishnan V M 1996 *Mater. Lett.* **29** 19
[6] Tamarin Y 2002 *Protective Coatings for Turbine Blades* (Ohio: Materials Park)
[7] Kianicová M 2006 *Fireproof diffusive coating degradation in consequence of short-time operating temperature exceeding* PhD Thesis (Trencin: Alexander Dubcek University)
[8] Kráľ J, Ferdinandy M, Kottfer D, Maňková I, Ceniga L, 2009 *Surface Rev. and Letters*, **16** 623
[9] Sozańska M, Chmiela B, Kianicová M and Cwajna J 2012 *Solids State Phen.* **186** 143
[10] Chmiela B, Sozańska M and Rodak K 2012 *Solid State Phen.* **186** 58
[11] Pokluda J and Kianicová M 2010 *Gas Turbines. Damage and Performance Aassessment of Protective Coatings on Turbine Blades* (India: Ed. Gurrappa Injeti. Sciyo.)

IOP Conf. Series: Materials Science and Engineering **35** (2012) 012011 doi:10.1088/1757-899X/35/1/012011

Changes of microstructure in titanium alloy during high temperature hydrogen treatment

M Sozańska
Department of Materials Science and Metallurgy, Silesian University of Technology, 40-019, Katowice, Poland

E-mail: maria.sozanska@polsl.pl

Abstract. The paper regarding the HTM - *Hydrogen Treatment of Materials* as an attractive method for controlling microstructure and improving final mechanical properties of titanium alloys. A new variant of high-temperature hydrogen treatment that includes cyclic heat treatment is proposed. This paper present the qualitative description of the microstructure in Ti-6Al-4V alloy after high-temperature hydrogen treatment with cyclic heat treatment operation. Hydrogen treatment consists of the three stages: hydrogenation in hydrogen gas atmosphere (600°C), cyclic heat hydrogen treatment (580°C), dehydrogenation in vacuum (550°C). The possibilities of hydrogen treatment for controlling microstructure layer of titanium alloy was presented.

1. Introduction

The positive impact of hydrogen on the properties of titanium alloys is manifested in the high temperature hydrogen treatment (HTM - Hydrogen Treatment of Materials), where hydrogen is a temporary alloying element. Titanium and its alloys show the solubility of almost 60 at.% of hydrogen. The limit solubility of hydrogen in Ti_{α} is very low and does not exceed 0.05 at.% at room temperature. The limit solubility of hydrogen in Ti_{β} is much higher and its maximum value is 48 at.%. In titanium alloys, hydrogen most often appears as hydrides. Most stable is δ hydride (TiH_x) with face centred cubic lattice (f.c.c.) and at room temperature it already appears at low concentrations of hydrogen [1-4].

Hydrogen treatment of materials (HTM) is a powerful method for both microstructure and properties modifications of two-phase titanium alloys (α + β). Hydrogen treatment causes grain refinement and therefore allows to achieve better general properties of these alloys. HTM consists of several stages and during these stages some structural transformations occur. In titanium alloys, hydrogen most often appears as hydrides. Specific volume of titanium hydrides is by 13÷17% higher in relation to specific volume of α phase, which causes large stress in crystal lattice of this phase and may result in its local plastic deformation [5,6]. High-temperature hydrogen treatment consists of the following four consecutive stages:

1. **First stage** – hydrogenation of titanium alloy at the temperature of two-phase area to produce rich in hydrogen the β_H - phase and the δ - phase (titanium hydride). Practice - holding at a temperature of two-phase α+β area in the atmosphere of hydrogen with defined pressure.
2. **Second stage** consists in allotropic change and the δ - phase decomposition at the temperature of the β - phase area. Practice - holding at a temperature of β phase area.

3. **Third stage** is annealing at the temperature of the (α + β) area. During this stage the transformation $\beta_H \rightarrow \alpha + \delta$ occurs. Practice - a long-term isothermal soaking at a temperature below the eutectoid transformation temperature.
4. **Fourth stage** – dehydrogenation in high vacuum at the temperature of the two-phase area which causes recrystallisation of the microstructure. After hydrogen treatment grain refinement and division of lamellar microstructure α - phase was observed (especially in the surface of specimens). Practice - an isothermal soaking at a temperature of two-phase α+β area.

A new variant of high-temperature hydrogen treatment that includes cyclic heat treatment is proposed. The idea of this treatment is based on accumulation after each consecutive cycle microstructure defecting as a result of execution of a single heating $\leftrightarrow$ cooling cycle. The reason for this defecting is internal-work hardening caused by interaction between individual phases representing the alloy microstructure as a result of diversity of their specific volumes. The combination of cyclic heat treatment operations, conducted at moderate heating and cooling rates, and recrystallisation annealing within the range of temperature of two-phase area may result in reduction in grain size of titanium alloys [8]. It consists of the following three consecutive stages (Fig.1):

1. **hydrogenation** – to obtain the microstructure that will consist of α phase, hydrogen-rich β_H phase and δ titanium hydride,
2. **cyclic heat treatment**: low-temperature treatment using eutectoid transformation and high-temperature treatment using eutectoid and allotropic transformations to obtain the appropriate microstructure defecting level as a result of an internal-work hardening caused by diversity of specific volumes of the phases occurring in the eutectoid and allotropic transformations that is repeated several times,
3. **dehydrogenation** – to remove hydrogen and, first of all, to forced recrystallisation of microstructure defected due to cyclic heat treatment resulting in grain refinement.

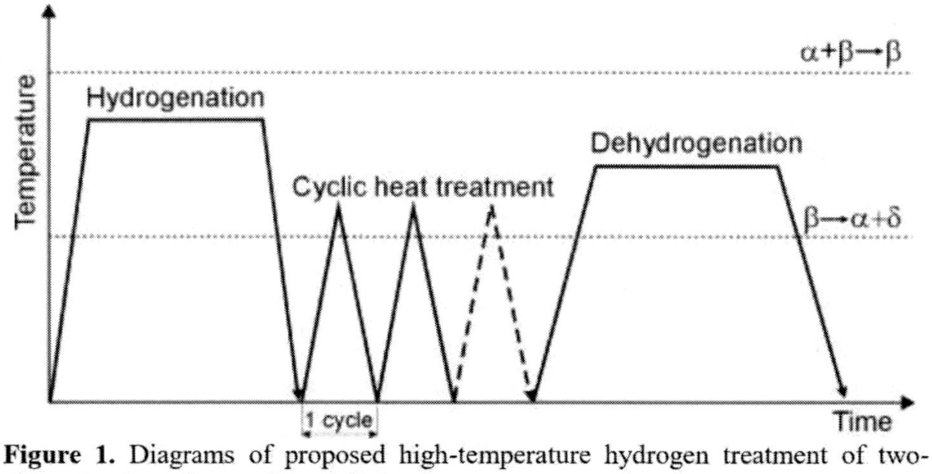

Figure 1. Diagrams of proposed high-temperature hydrogen treatment of two-phase titanium alloys using cyclic heat treatment.

The proposed variant of high-temperature hydrogen treatment, including the proprietary ones that use cyclic heat treatment (Fig. 1), should make it possible to grain refinement and improve strength and plastic properties as well as technological plasticity of the previously hydrogenated two-phase titanium alloys with coarse-grained initial microstructure by using heat treatments only, without plastic deformation.

The aim of this study was to determine the effect of hydrogen on the two-phase microstructure, hardness of titanium alloy Ti-6Al-4V after high temperature hydrogen treatment.

2. Material and experimental procedure

Cylindrical specimens a diameter of 16 mm and a height of 12 mm made of titanium Ti-6Al-4V alloy. The chemical composition (% mass.) was: Al- 6.2, V- 4.3, Fe- 0.3, and Ti-the rest. Annealing was performed at 1100 °C for 1 hour, subsequent cooling of the furnace. Hydrogenation processes and cyclic treatment after hydrogenation conducted on a universal device for machining a glow from the cold chamber walls. Process parameters used were as (Fig. 2):
Step 1: *hydrogenation* - soaking 60 min. at 600 °C,
Step 2: *cyclic heat treatment* - at 580 °C cooling to a temperature of 250 ° C and soak for 30 minutes, number of cycles - 10, gas atmosphere of 100% H_2,
Step 3: *dehydrogenation* - soaking 60 min. at 550 ° C in a vacuum, and *cooling* after hydrogenation in a vacuum.

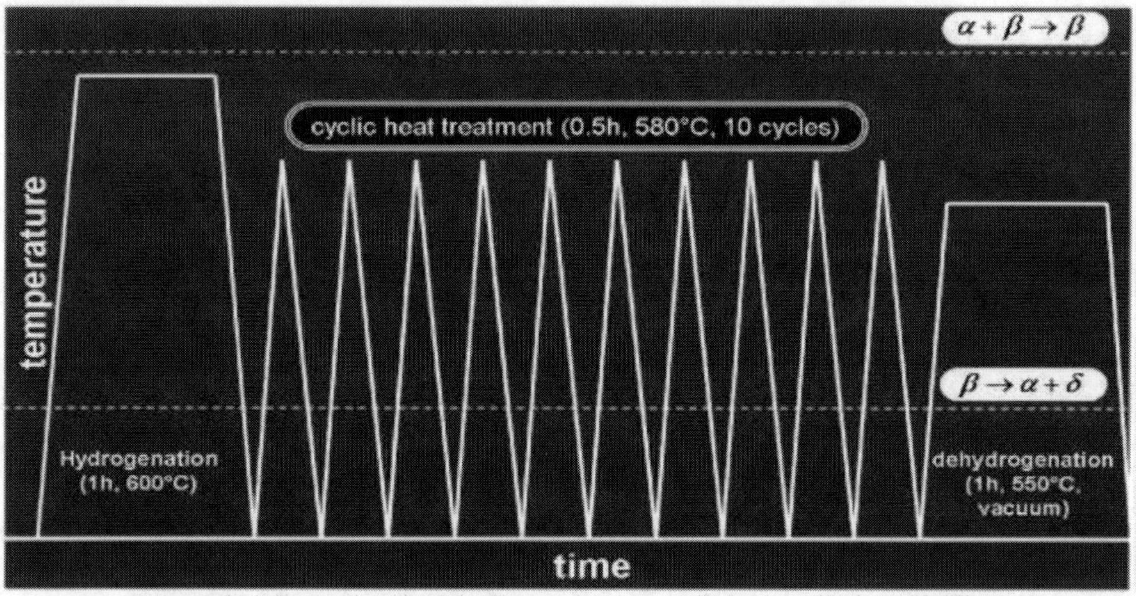

Figure 2. Schematic of high hydrogen treatment of titanium Ti-6Al-4V alloy.

The specimens surface was etched in a two solutions: I solution - of 30 cm^3 of lactic acid, 15 cm^3 HNO_3, 5 cm^3 HF to reveal grains and II solution – of 50 cm^3 of glycerol, 25 cm^3 HNO_3, 30 cm^3 H_2O, 1 cm^3 HF for metallographic examination.
A macrostructural examination of the specimens surface was performed using a stereoscopic microscope (Olympus SZX-9). The microstructure was characterized using a light microscopy OLYMPUS GX71 and scanning electron microscopy SEM (Hitachi S-3400N). Tests of hardness HV10 was carried out on automated Struers Duramin A300 (HV10).

3. Results and discussion

3.1. Effect of high temperature hydrogen treatment on the microstructure of Ti-6Al-4V alloy

Figure 3 shows the specimens macrostructure of Ti-6Al-4V alloy before and after high hydrogen treatment (HTM). Some difference in the grains size are visible.

a)

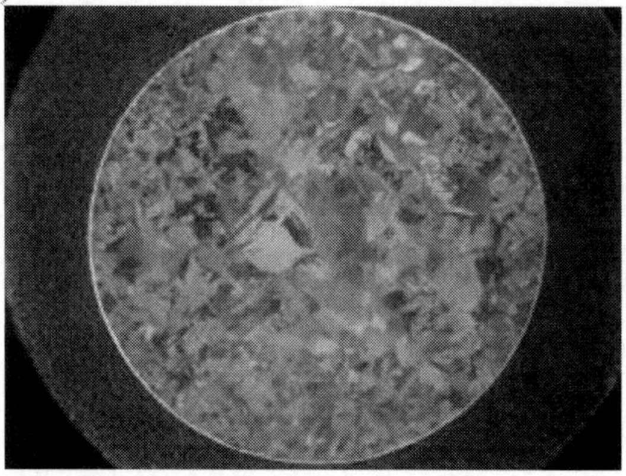

b)

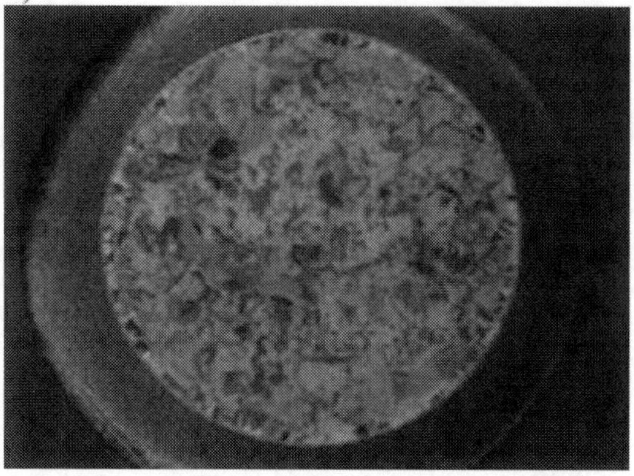

Figure 3. Macrostructure of titanium Ti-6Al-4V alloy: a) before HTM, b) after HTM.

The alloy investigated in the initial state after full annealing within the β single-phase area temperature were characterised by homogeneous two-phase microstructure. The alloy microstructure is represented by alternately arranged lamellar precipitations of α and β phases with different orientations in particular grains (Figs. 4 and 5). An SEM micrograph of the cross-section reveals the microstructure is comparably in center and edge specimens before HTM (Fig. 6).

a)

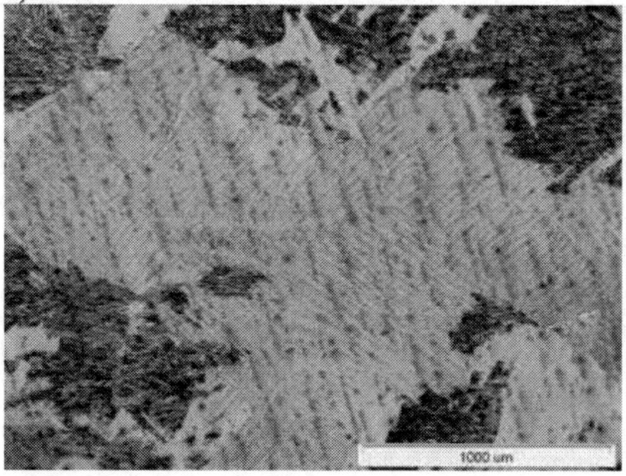

b)

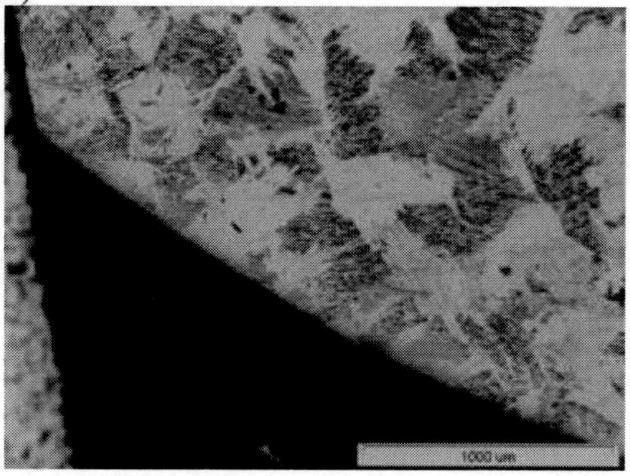

Figure 4. Microstructure of titanium Ti-6Al-4V alloy before HTM: a) center specimen, b) edge specimen, LM

IOP Conf. Series: Materials Science and Engineering **35** (2012) 012011 doi:10.1088/1757-899X/35/1/012011

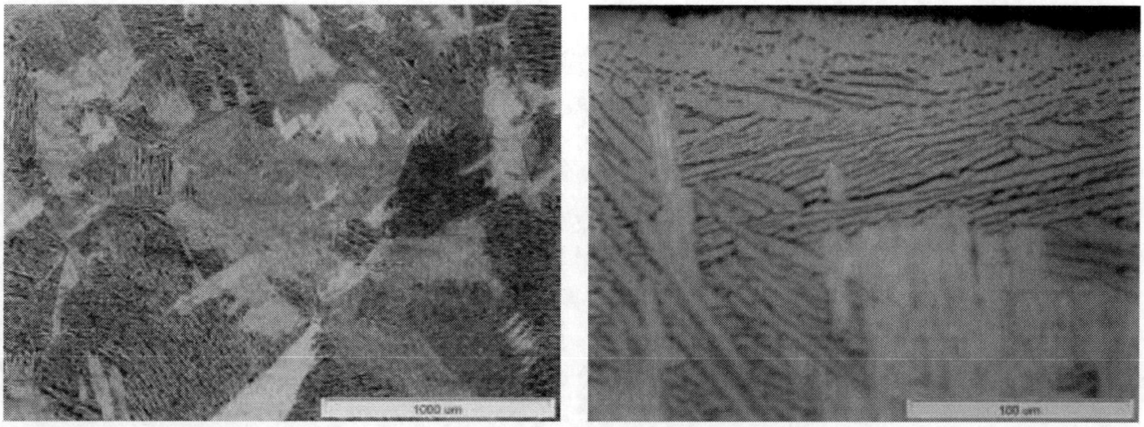

Figure 5. Microstructure of titanium Ti-6Al-4V alloy after HTM: a) center specimen, b) edge specimen, (10 cycle), LM.

a) b)

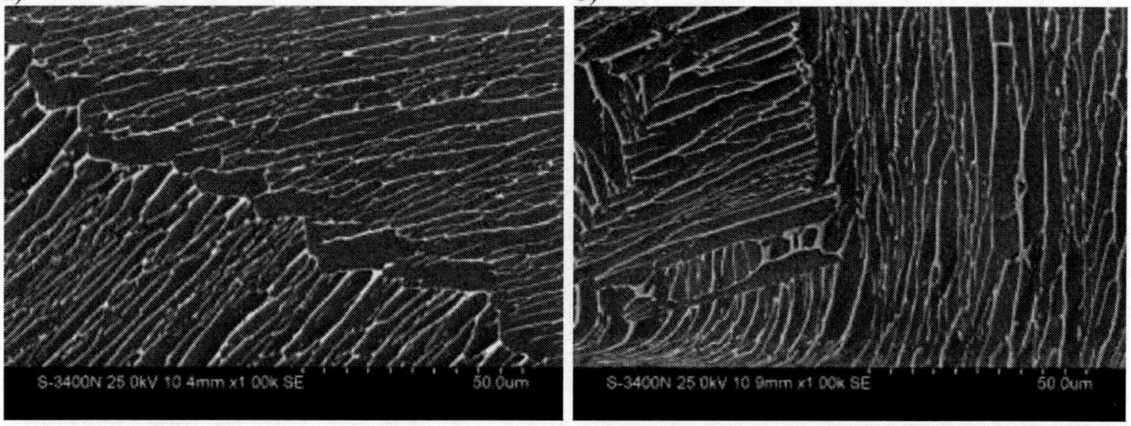

Figure 6. Microstructure of titanium Ti-6Al-4V alloy before HTM: a) center specimen, b) edge specimen, SEM

Analysis of microstructural changes in the investigated alloys after high-temperature hydrogen treatment with cyclic operation shows different changes in microstructure in the centre and in the edge part of specimen (Fig. 7). In the centre, not significant changes in the microstructure were observed (Fig. 7 a,b). Moreover, in the edge part of specimen the lamellar precipitations of α phases were divided and the α phase relative volume was reduced (Fig. 7 c,d),

High-temperature hydrogen treatment led to the fragmentation of lamellar precipitations of the phase. The α phase dissolution process was limited for 50μm (Fig. 8 a,b). Analysis of the edge of specimen in Ti-6Al-4V alloy after high-temperature hydrogen treatment with 10 cycles hydrogen treatment shows a lot of cracks (Fig. 9 a,b).

IOP Conf. Series: Materials Science and Engineering **35** (2012) 012011 doi:10.1088/1757-899X/35/1/012011

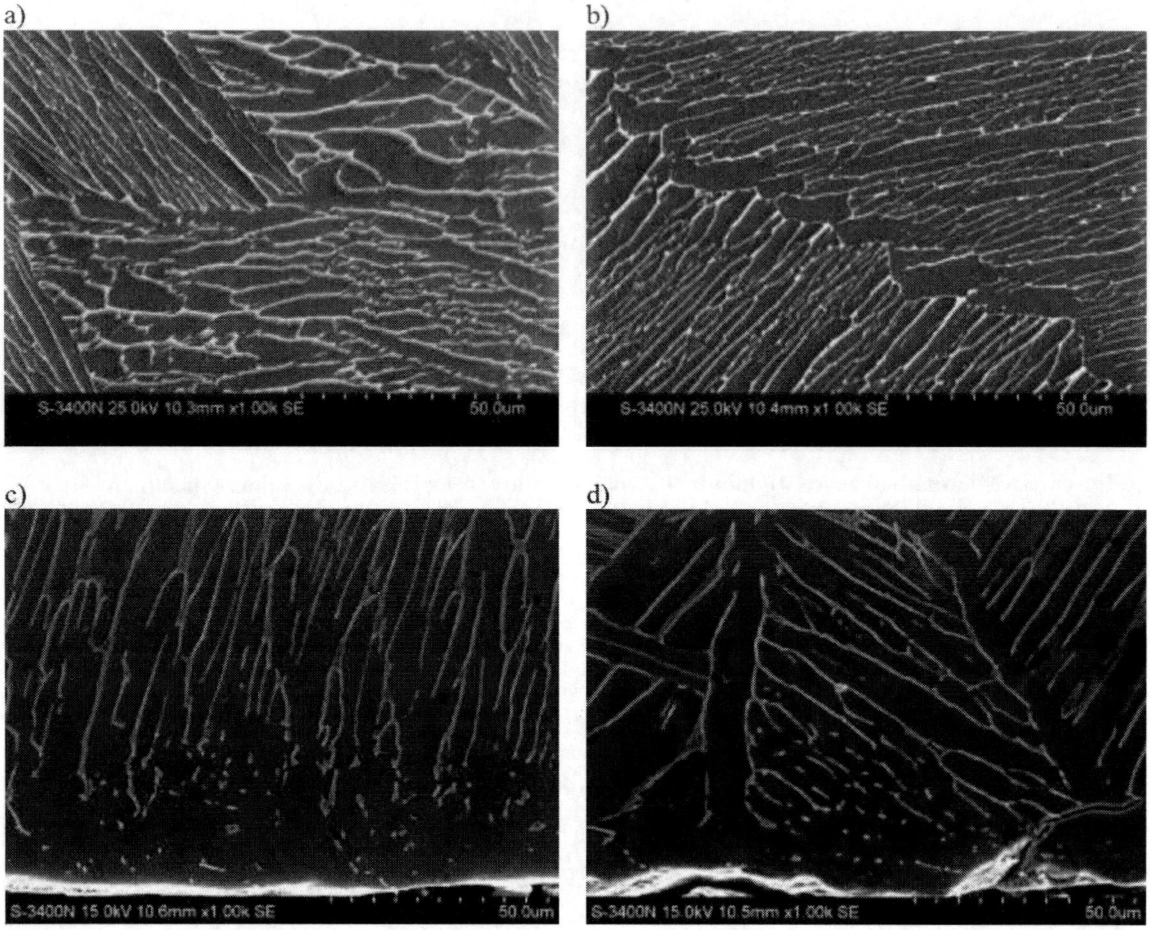

Figure 7. Microstructure of titanium Ti-6Al-4V alloy after HTM: a, b) center specimens, c, d) edge specimen, (10 cycle), SEM.

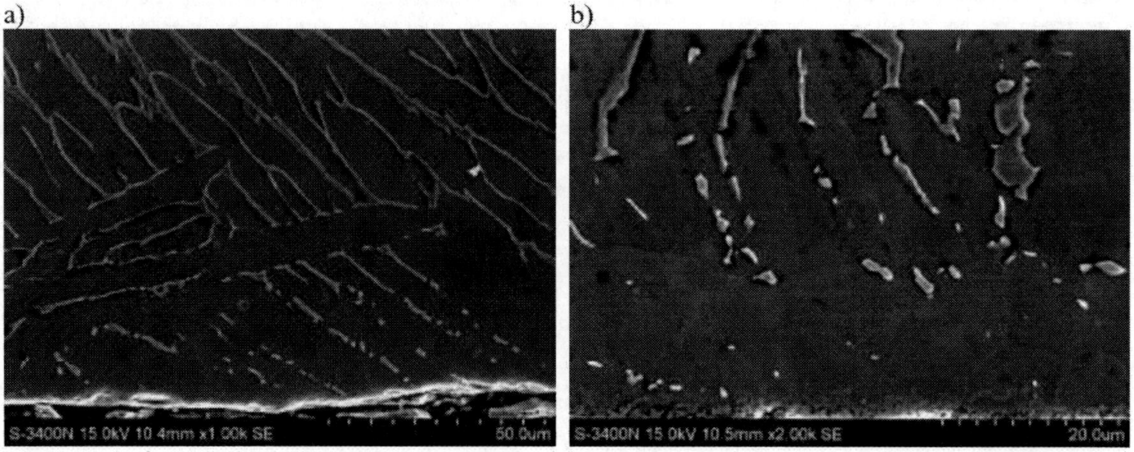

Figure 8. Microstructure of titanium Ti-6Al-4V alloy after HTM: edge specimens, (10 cycle), SEM.

a) b)

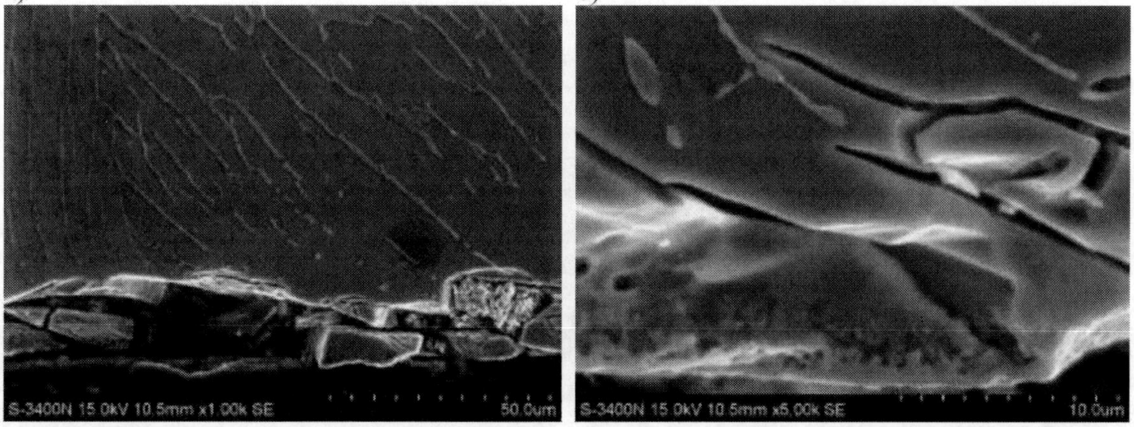

Figure 9. Cracks in microstructure of titanium Ti-6Al-4V alloy after HTM: edge specimens, (10 cycle), SEM.

3.2. Effect of high temperature hydrogen treatment on hardness of Ti-6Al-4V alloy
HV10 hardness results before and after high temperature hydrogen are shown in Table 1.

Table 1. The results of hardness before and after HTM.

Hardness / **Treatment**	**Center,** HV10	**Egde,** HV10
Ti-6Al-4V alloy before HTM	**314** ± 7.09	**358** ± 29.34
Ti-6Al-4V alloy after HTM, 10 cycle	**322** ± 15.13	**422** ± 17.28

Analysis of the results of hardness (Table 1) indicates significant differences between the results for the sample before HTM and after HTM. Both, in sample before and after HTM, the hardness is higher than the edge part than in the center, but these differences are in the range of measurement error. The results of hardness of the hydrogen-treated samples according to 10 cycle indicate that the HV10 hardness values in the layer are larger than in the center of the samples.

4. Conclusions
High-temperature hydrogen treatment with a cyclic heat temperature hydrogen treatment is a very powerful method for microstructure modification of the titanium alloy. Especially, HTM is a very efficient method for changing the microstructure and the hardness of the layers of two-phase titanium alloys.

The scheme of phase transformation in high temperature hydrogen treatment of titanium Ti-6Al-4V alloy is shown in figure 10.

IOP Conf. Series: Materials Science and Engineering **35** (2012) 012011 doi:10.1088/1757-899X/35/1/012011

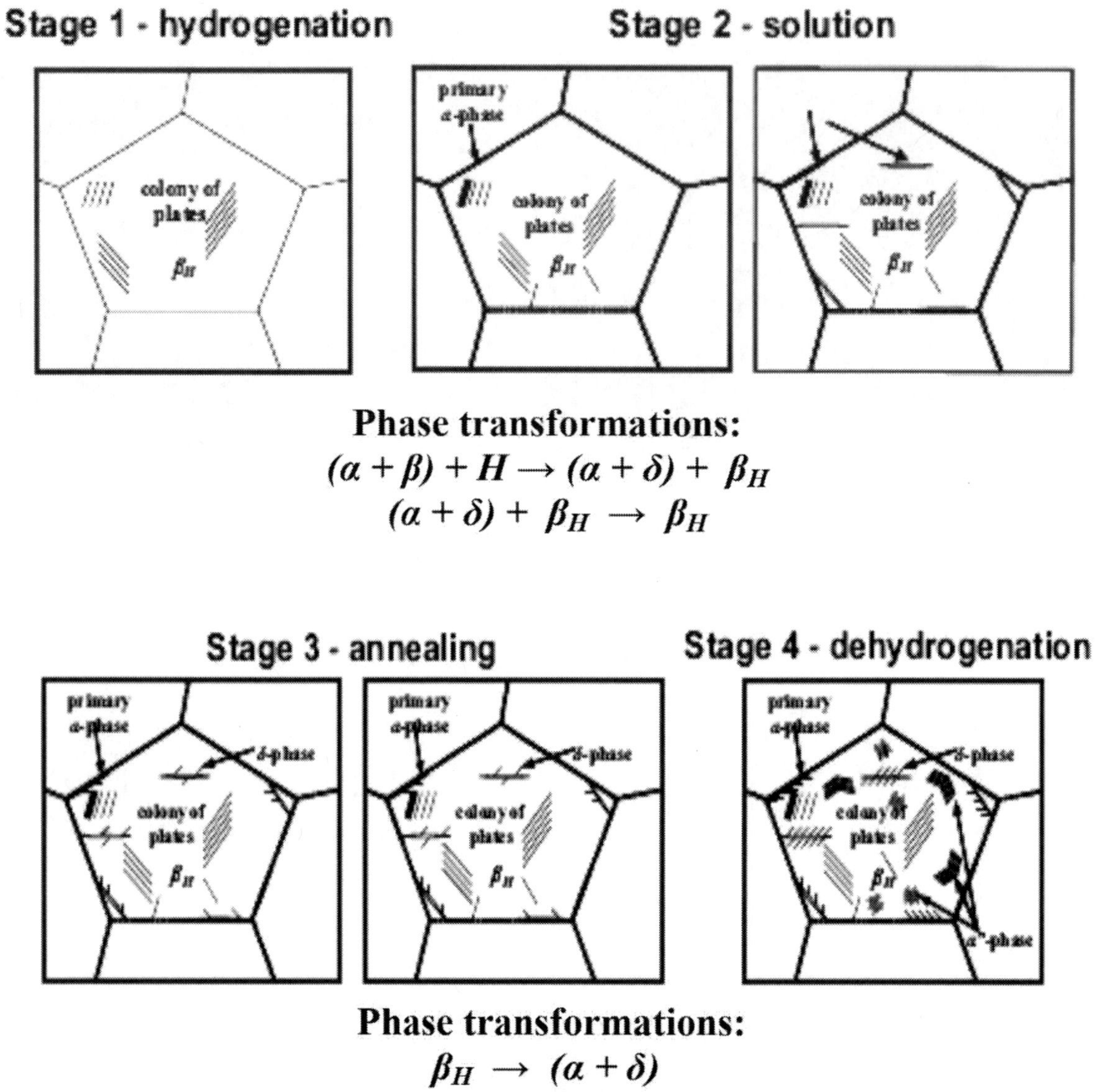

Figure 10. Scheme of phase transformation in hydrogen treatment of titanium Ti-6Al-4V alloy.

Analysis of research on effects of hydrogen on the microstructure of titanium Ti-6Al-4V alloy showed that the processes causing the high temperature hydrogen treatment of grain refinement and change the structure in layer up to 50 μm. For all samples, irrespective of the applied high-temperature variant of the hydrogen treatment were found differences between the microstructure in the center and the edge regions. Plates located in the layer have been markedly fragmented, while in the center of the samples changes were not observed. Microcraks were detected after all the processes hydrogen treatment. High temperature hydrogen treatment changes on the hardness of Ti-6Al-4V alloy.

Acknowledgements
This scientific work is financed from the budget funds for science in the years 2009-2012 as the research project of the Ministry of Scientific Research and Information Technology No. N N507 462337.

References
[1] Tkachov V 2000 *Mat. Sci.* **36** No 4 215.
[2] Zieliński A 1999 *Niszczenie wodorowe metali nieżelaznych i ich stopów*, Gdańskie Towarzystwo Naukowe, Gdańsk (in polish).
[3] Froes F H, Eliezer D, Senkov O N and Quazi J L: *Beneficial effect as a temporary alloying element in titanium alloys: An overview*, "Hydrogen Effect on Materials Behavior and Corrosion Deformation Interactions", Ed. by N.R. Moody, A.W. Thompson, R.E. Ricker, G.W. Was and R.H. Jones, TMS, 301-314.
[4] Qazi J L, Rahim J, Senkov O N, Patankar S N, Froes F H 2003 *Phase transformations in the Ti-6Al-4V – hydrogen system*, "Hydrogen Effect on Materials Behavior and Corrosion Deformation Interactions", Ed. by N.R. Moody, A.W. Thompson, R.E. Ricker, G.W. Was and R.H. Jones, TMS (The Minerals, Metals & Materials Society), 337-344.
[5] Senkov ON et al. 2001 *Mater. Res. Bull.* **36** 1431.
[6] Kaibyshev O A 2001 *J. Mater. Proces. Technol.* **117** 300.
[7] Simka W, Sadkowski A, Warczak M, Iwaniak A, Dercz G, Michalska J and Maciej A 2011 *Electrochim. Acta* **56** 8962.
[8] Szkliniarz W 2002 *Inżynieria Materiałowa* **3** 96 (in polish).

IOP Conf. Series: Materials Science and Engineering **35** (2012) 012012 doi:10.1088/1757-899X/35/1/012012

Hydrogen effects in duplex stainless steel welded joints – electrochemical studies

J Michalska[1], J Łabanowski[2], J Ćwiek[3]
[1]Department of Materials Science, Silesian University of Technology, Krasinskiego 8, 40-019 Katowice, Poland
[2]Department of Materials and Welding Engineering, Gdansk University of Technology, G. Narutowicza 11/12, 80-233 Gdańsk, Poland
[3]Institute of Engineering Materials and Biomaterials, Silesian University of Technology, Konarskiego 18A, 44-100 Gliwice, Poland

E-mail: joanna.k.michalska@polsl.pl

Abstract. In this work results on the influence of hydrogen on passivity and corrosion resistance of 2205 duplex stainless steel (DSS) welded joints are described. The results were discussed by taking into account three different areas on the welded joint: weld metal (WM), heat-affected zone (HAZ) and parent metal. The corrosion resistance was qualified with the polarization curves registered in a synthetic sea water. The conclusion is that, hydrogen may seriously deteriorate the passive film stability and corrosion resistance to pitting of 2205 DSS welded joints. The presence of hydrogen in passive films increases corrosion current density and decreases the potential of the film breakdown. It was also found that degree of susceptibility to hydrogen degradation was dependent on the hydrogen charging conditions. WM region has been revealed as the most sensitive to hydrogen action.

1. Introduction

Duplex stainless steels (DSS) are modern structural materials with excellent corrosion resistance and enhanced mechanical properties which have been developed as an attractive techno-economical substitute to austenitic grades. A phase-balanced and defect free microstructure is critical for desirable properties. Welding processing, if not carried out with careful control over welding parameters and consumables, can create phase imbalance and formation of undesirable phases [1,2]. Due to the rapid cooling rate experienced by the welded joint, excessive ferritization can occur, while upon slow cooling precipitation of various intermetallic phases can occur in the heat-affected zone (HAZ) and weld metal (WM) of DSS [3].

Inspection programmes indicate that DSS may suffer from hydrogen and the effects of hydrogen may be enhanced in DSS welded joints. The influence of hydrogen on the microstructure [4] and mechanical properties [5,6] of DSS are well documented. In spite of a number of research performed on hydrogen embrittlement of DSS, it is uncertain which factors are crucial in the development of hydrogen degradation. Interactions of hydrogen in passive films may enhanced the susceptibility to environmental degradation of stainless steels, because the entering into steel hydrogen affects its electrochemical properties. Therefore, the understanding and control of passivity in hydrogen charged material has been a key factor for protection of stainless steels against corrosion. A few researchers have reported the effects of cathodically evolved hydrogen on localized corrosion in stainless steels [7-9]. However, the effect of hydrogen on electrochemical behavior of DSS welds so far has not been investigated. Considering the facts that welding is an integral part of most fabrication processes of stainless steel components and that the possibility of hydrogenation may contribute to premature damage of DSS constructions , the assessment of corrosion resistance of DSS welded joints is absolutely essential. The present work has been aimed at further claryfing the effect of hydrogenation on the corrosion behaviour and passivity of 2205 DSS welded joints. Three different zones from the welded joints (WM, HAZ an parent metal) were taking into account within electrochemical measurements.

2. Experimental procedure

2.1. Material

The plate of 15 mm thickness made of 2205 DSS in as-received condition was used for the study. The as-received material was solution annealed at 1100 °C for 0.5 h to homogenize the microstructure. The nominal chemical composition of investigated steel is presented in Table 1. The value of the pitting resistance equivalent (PRE) calculated from the following expression: wt%Cr + 3.3(wt%Mo) + 16(wt%N), corresponds to 23.

Table 1. Chemical composition of 2205 duplex stainless steel.

Content of chemical elements (wt. %)									
C	Cr	Ni	Mo	Mn	Si	P	S	N	Fe
0.021	22.70	5.95	3.20	0.88	0.39	0.020	0,001	0.16	balance

2.2. Welding parameters

Butt joints on the plates 15 mm in thickness were performed with the use of SAW method. The parameters used for welding process are listed in Table 2. Welded joints were performed using 2Y square edge preparation. Duplex stainless steel filler metal 22Cr-9Ni-3Mo and basic non-alloyed agglomerated flux (ESAB Flux 10.93) were used. Two beds were performed to fill the whole joint with the use of heat inputs not exceed 2.5 kJ/mm. The interpass temperature was limited to 100 °C maximum. Each weld was X-rayed and crack tested, and found to be satisfactory with B quality class according to EN 25817 standard. The macrostructure of the welded joint is presented in figure 1.

Table 2. Parameters used for weldment preparation.

Parameter	Quantity
Current	600 A
Current polarity	DC+
Voltage	33 V
Speed	6.7 mm/s

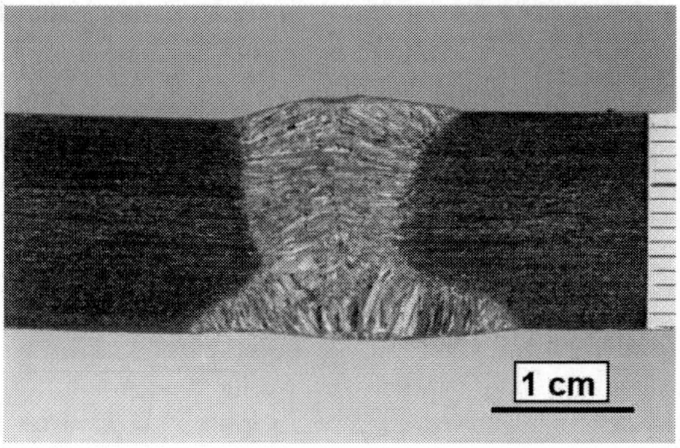

Figure 1. Macrostructure of 2205 DSS welded joint.

2.3. Metallographic examinations

Metallographic examinations were aimed to determine the general microstructure of the welded metal and heat affected zone. Microscopic observations were performed to find the presence of secondary austenite and precipitations of any intermetallic phases. Phases were identified by the color contrast produced by electrolytic etching the in 10 N KOH solution at 3 V.

Figure 2 shows microstructures of the welded joint showing WM, HAZ and parent metal (PM). Structure of weld metal on the whole cross section is similar. During solidification of duplex weld metal an almost completely ferrite structure is formed. Further cooling initiates the formation of the austenite phase nucleating at the ferrite grain boundaries. In examined welds a dendritic austenite microstructure developed in fast cooling conditions (Fig. 2b). More globular microstructure was detected at the root run due to the presence of secondary austenite.

Heat affected zone microstructure could be critical for welded joint properties. For examined welds the very narrow zones of about 300-500 μm were observed (Fig. 2a). The ferrite content in that zone was significantly higher in comparison to bulk weld metal microstructure. The HAZ microstructure consists of lamellar austenite precipitates located mainly on equiaxial great ferrite grain boundaries. There was no evidence of secondary phases in HAZ microstructure

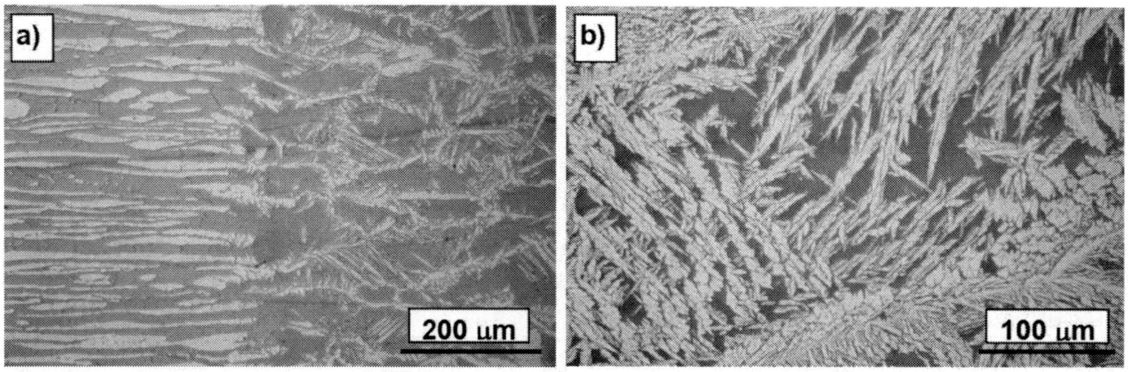

Figure 2. Microstructure (LM) of welded 2205 DSS showing: (a) parent metal (PM), HAZ areas (b) magnified WM area.

2.4. Specimen Preparation

Coupons of 10 mm x 10 mm size were cut from the welded DSS plate contained WM, HAZ, and parent metal. Grinding was carried out using silicon carbide emery paper to a grit size of 1200 followed by three step diamond polishing up to 1 μm finish. Degreasing of the polished coupons was carried out by ultrasonic cleaning in ethanol for 5 minutes.

2.3. Hydrogen charging

Hydrogen was introduced into the samples by cathodic current method under galvanostatic condition at room temperature. The charging solution was 0.1 N NaOH + 1 mg dm^{-3} As_2O_3. Basic solution (0.1 N NaOH) was used to prevent corrosion damage and to avoid contamination of the sample surface during hydrogen charging. During the cathodic polarization, elemental arsenic created from As_2O_3 is reduced to AsH_3, which is the most effective promoter of hydrogen entry into a steel [10]. The coupons contained WM, HAZ and parent metal regions were serially charged at current densities of 1, 2 and 10 mA cm^{-2} for 1 week, respectively. The increase in cathodic current density caused the increase in hydrogen absorption level.

2.4. Electrochemical measurements

A potentiodynamic polarization technique was used for corrosion studies. Polarization curves were recorded on both uncharged and hydrogen charged samples in a synthetic sea water type A solution at room temperature. The investigations were conducted using a measuring system composed of a computer controlled Solartron 1287 electrochemical interface and traditional three-electrode flat type cell, wherein the saturated calomel electrode (SCE) acted as a reference electrode and a platinum wire as a counter electrode. The surface area during measurement was 0.2 cm^2. The obtained results were elaborated by means of a Corrview (Scribner Inc.) software. The hydrogen charged samples, prepared as described earlier, were immediately transferred from the charging cell to the polarization cell for electrochemical studies. The measuring procedure started with recording the open circuit potential values (E_{OCP}) for 30 minutes. The electrode potential was anodically scanned at a rate of 0.167 mV/s from -300 mV vs. E_{OCP} . The scan was reversed at a current density of 0.001 A cm^{-2}. Corrosion current density i_{corr} and corrosion potential E_{cor} were established from linear polarization (Rp) measurements (± 50 mV vs. OCP). Three pitting parameters were determined from a potentiodynamic polarization curves: breakdown potential E_{pit}, repassivation potential E_{pit} and ΔE - the difference between pitting potential E_{pit} and repassivation potential E_{rep}.

3. Results and discussion

The results of linear polarization measurements in a synthetic sea water are listed in table 3. The analysis of obtained results revealed that corrosion parameters of 2205 DSS welded joints is influenced by hydrogen. E_{cor} values of hydrogen charged coupons were more active in comparison to E_{corr} for uncharged ones. They were found to be more cathodic with increase in hydrogen charging current. Increased corrosion current densities i_{corr} for hydrogen charged coupons were also observed.
The effect of hydrogen depends on the type of region on the welded joint. WM regions have been as the most sensitive to hydrogenation and hydrogen induce damage. It was found that as early as 1 mA cm^{-2} of hydrogen charging caused a twofold increase of i_{corr} compared to the corresponding coupon with parent metal. After hydrogenation at 10 mA cm $^{-2}$, the strongest effect was observed – value for WM was almost eighteen times higher. Coupons with HAZ regions were less prone to hydrogen damage. Gradual increase in i_{corr} was observed with an increase of hydrogen charging current. The maximal value obtained was 7.101 μA cm^{-2}.

Table 3. The results of linear polarization measurements.

Sample			E_{OCP} (mV)	E_{corr} (mV)	i_{corr} (μA cm^{-2})	R_p (kΩ cm^{-2})
uncharged		PM	31	133	0.029	899.4
		WM	-23	50	0.065	398.8
		HAZ	-68	-40	0.203	128.1
charged	1 mA cm^{-2}	PM	-312	-64	1.711	151.9
		WM	-268	-96	3.387	7.7
		HAZ	-424	-184	2.021	12.9
	2 mA cm^{-2}	PM	-298	-59	2.380	10.9
		WM	-322	-121	13.93	1.9
		HAZ	-410	-243	6.941	3.7
	10 mA cm^{-2}	PM	-542	-382	3.161	8.2
		WM	-651	-481	30.32	0.9
		HAZ	-670	-497	7.101	4.1

Similar observations could be performed evaluating passivation and pitting parameters. Figures 3-6 show complete polarization curves of coupons investigated. Figure 3 corresponds to uncharged coupons. The curves are typical for strongly passivated material: the anodic current density maintains a value of ≈ 1μA/cm^2 until a potential ~ 1 V, then an abrupt increase of the current density is observed. E_{pit} values are highly anodic and corresponding E_{rep} values are very close to E_{pit} with a small size of hysteresis loop. Situation has changed with the presence of hydrogen. Polarization curves illustrating the effect of hydrogen on the electrochemical behaviour of investigated coupons are shown in figures 4-6. An increase in passive current density was observed after hydrogenation of steel. In addition, the higher current of hydrogen charging, the higher passive currents were registered on a polarization curve. An increase in passive current densities indicated that the stability of passive film was affected.

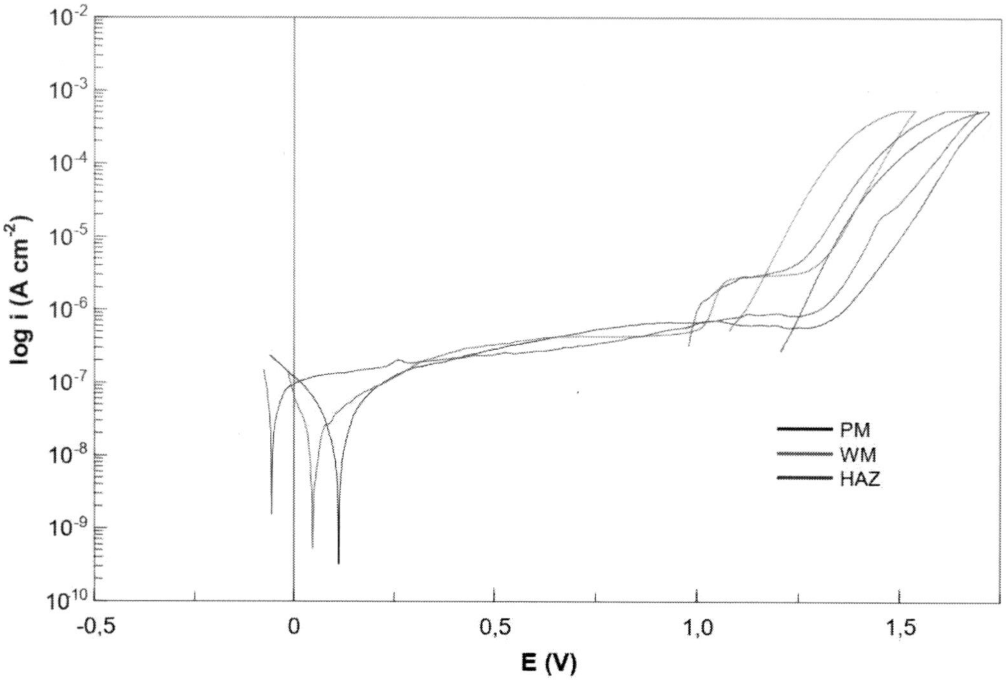

Figure 3. Polarization curves of 2205 DSS welded joint in synthetic sea water solution (coupons without hydrogen).

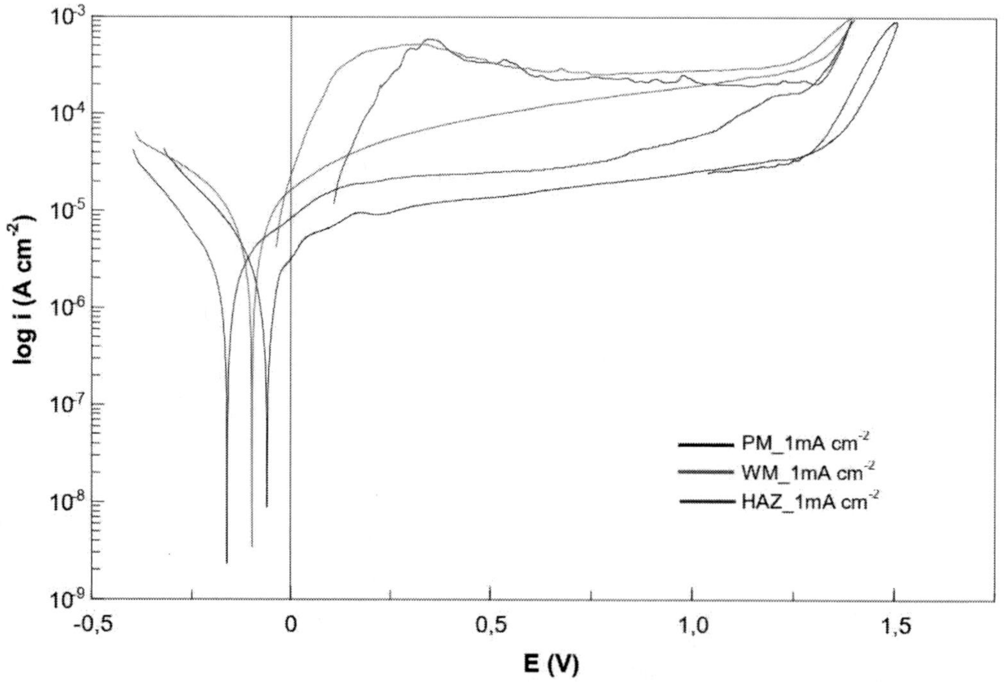

Figure 4. Polarization curves of 2205 DSS welded joint in synthetic sea water solution (coupons hydrogen charged at 1 mA cm^{-2}).

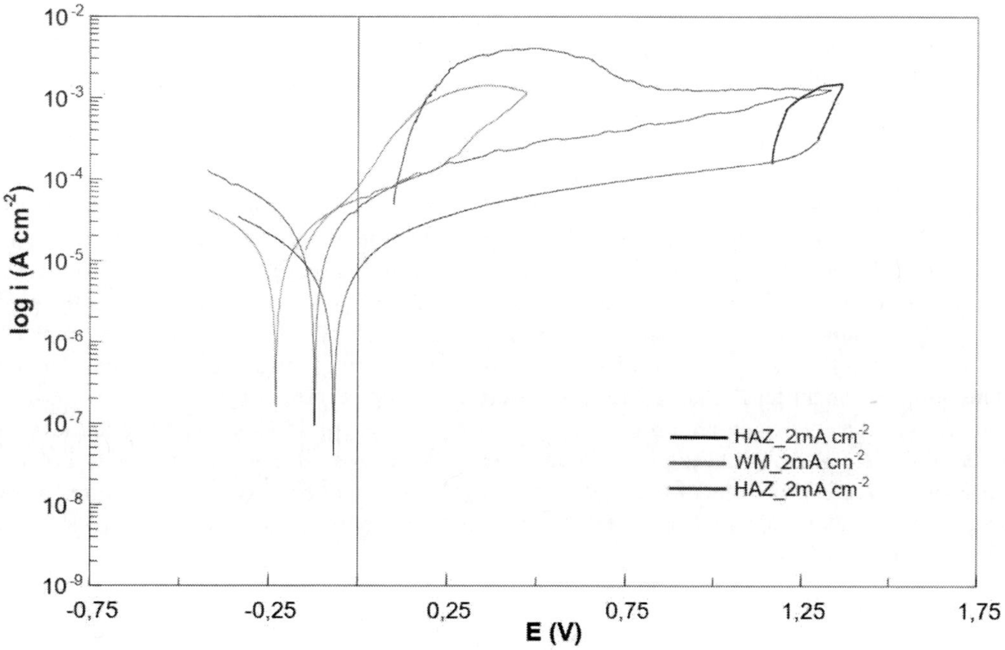

Figure 5. Polarization curves of 2205 DSS welded joint in synthetic sea water solution (coupons hydrogen charged at 2 mA cm^{-2}).

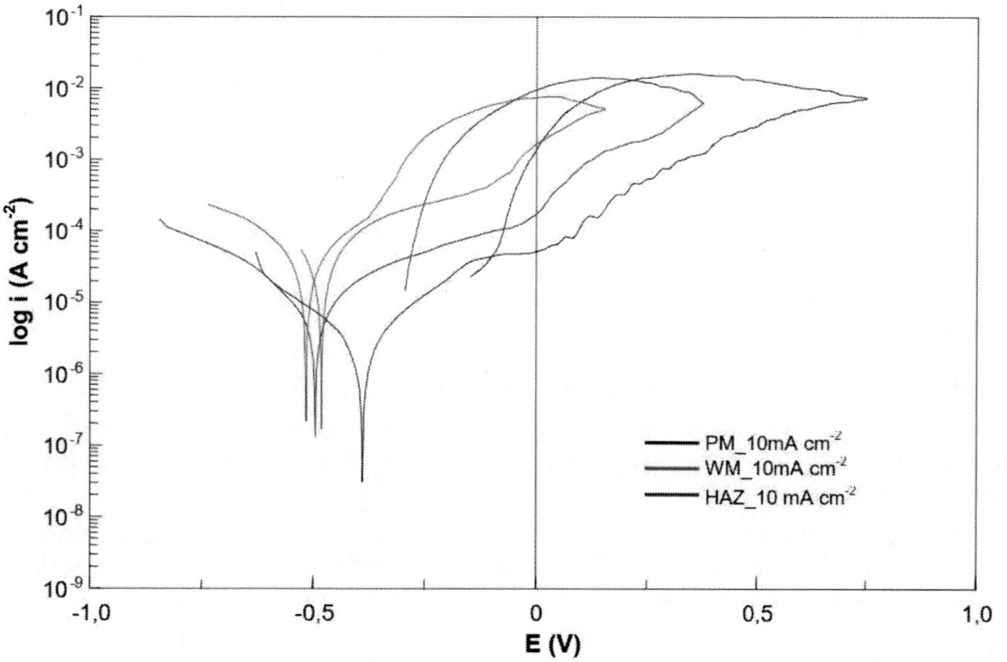

Figure 6. Polarization curves of 2205 DSS welded joint in synthetic sea water solution (coupons hydrogen charged at 10 mA cm^{-2}).

Hydrogen may seriously disturb and hinder passivation process on DSS and a protective layer is then weaker and more defected. Changes in the passive range had resulted in the shift of E_{pit} to less noble values, which unequivocally indicates that hydrogen may promote pitting on DSS and it has been revealed that this effect strongly depends on the microstructure and type of the region on welded joint. These relations are illustrated in figure 7.

Parent metal has been revealed as the most resistant area to pitting. Up to the charging current of 2 mA cm^{-2}, the width of passive range exceeds 1 V and breakdown of passive layer (E_{pit}) occurred above 1.25 V assuring high corrosion resistance. Very similar relations related to E_{pit} values were found for HAZ region. However, repassivation parameters for HAZ were strongly affected by hydrogen, even after a hydrogenation at 1 mA cm^{-2} pointing the lower resistance to pitting compared to the PM coupons. WM area has been revealed as the least resistant to pitting. Both, pitting and repassivation parameters were distinctly decreased by hydrogen. E_{rep} values were in the cathodic region even after hydrogenation 1 mA cm^{-2} suggesting strong effect of hydrogen on corrosion resistant and permanent loss of ability to repassivation.

Generally, after hydrogen charging at 10 mA cm^{-2}, the complete lack of resistance to pitting was observed for all coupons investigated. Breakdown of passive film occurred much earlier on the polarization curves irrespective of the type of a region on the welded joint. All E_{pit} values were shifted above 1 V into the more active direction compared to the uncharged coupons and E_{rep} parameters were negative, even for PM.

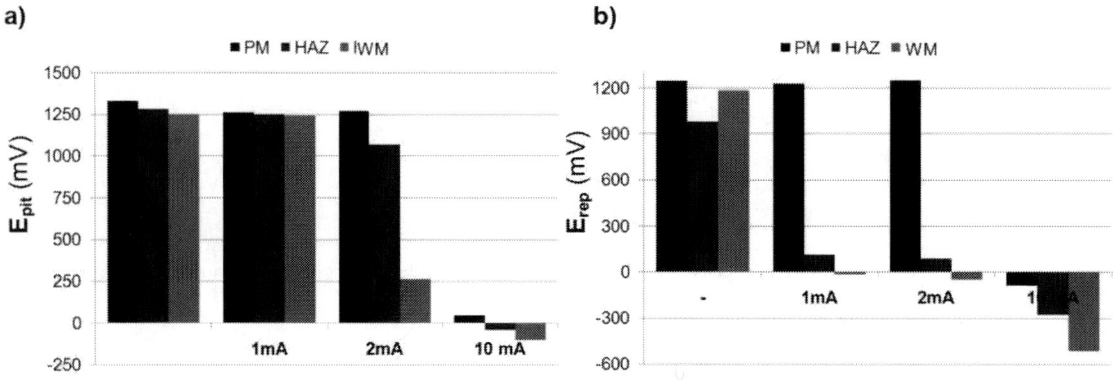

Figure 7. The effect of hydrogen charging conditions on: (a) pitting potential, (b) repassivation potential in 2205 DSS welded joints.

4. Conclusions

The results of the present investigation showed that hydrogen affects the corrosion resistance of 2205 DSS welded joints. The following conclusions can be drawn from this study:

1. The effect of hydrogen depends on the microstructure and type of the region on welded joint. WM area has been found as the most sensitive to hydrogen damage, while PM the most corrosively resistant to hydrogen action.
2. Cathodic hydrogen may strongly influenced the corrosion resistance of 2205 DSS welded joints. The presence of hydrogen shifted E_{corr} to the more active direction and cause an increase i_{corr} values. The higher current of hydrogen charging, the lower corrosion resistance.
3. The presence of hydrogen may seriously disturb and hinder passivation processes on DSS welded joints resulting in decreased resistance to pitting. This effect was particularly emphasized in repassivation tendency.

Acknowledgements
Scientific work was supported by the Polish Ministry of Science and Higher Education in the years 2010-2012 under the research project No. IP 2010 025870.

IOP Conf. Series: Materials Science and Engineering **35** (2012) 012012 doi:10.1088/1757-899X/35/1/012012

References:

[1] Tavares S S M, Pardal J M, Lima L D, Bastos I N, Nascimento A M and de Souza J A 2007 *Mater. Charact.* **58** 610.
[2] Gooch T G 1991 in Charles J and Bernhardsson N(eds.) *Duplex Stainless Steels 91*, Les Editions de Physique, Les Ulis, France, 211–15.
[3] Varol I, Baeslack W A and Lippold J C 1989 *Metallography* **23** 1.
[4] Sozanska M, Cwajna J and Sojka J 2004 *Acta Metall. Slovaca* **1** 798
[5] Luu W C, Liu P W and Wu J K 2002 *Corr. Sci.* **44** 1783.
[6] Chen S S, Wu J K and Wu T I 2004 *J. Mater. Sci.* **39** 67
[7] Ningshen S, Kamachi Mudali U, Amarendra G, Gopalan P, Dayal R K and Khatak H S 2006 *Corrosion Science* **48** 1106.
[8] Yang Q and Luo J L 2000 *Thin Solid Films* **371** 132.
[9] Michalska J 2011 IOP Conf. Ser. Mater. Sci. Eng. **22** 012016.
[10] Zakroczymski T 1991 *Electrochemical aspects of hydrogen entry into iron and steel from aqueous solutions* in Flis J (Ed.) Corrosion of Metals and Hydrogen-Related Phenomena: Selected Topics, (Elsevier)

IOP Conf. Series: Materials Science and Engineering **35** (2012) 012013 doi:10.1088/1757-899X/35/1/012013

Hydrogen assisted cracking of 2205 duplex stainless steel in synthetic sea water

J Michalska[1], J Łabanowski[2], J Ćwiek[3]
[1]Department of Materials Science, Silesian University of Technology, Krasinskiego 8, 40-019 Katowice, Poland
[2]Department of Materials and Welding Engineering, Gdansk University of Technology, G. Narutowicza 11/12, 80-233 Gdańsk, Poland
[3]Institute of Engineering Materials and Biomaterials, Silesian University of Technology, Konarskiego 18A, 44-100 Gliwice, Poland

E-mail: joanna.k.michalska@polsl.pl

Abstract. The cracking behavior of 2205 duplex stainless steels (DSSs) in synthetic sea water under cathodic polarization condition was investigated. (SSRT) method was employed in aim to evaluate the susceptibility to hydrogen assisted cracking. The results showed that the reduction in the uniform elongation (UEL) and the reduction of area (RA) varied with the applied cathodic current density. Significant reductions in ductility were found, indicating its susceptibility to hydrogen-assisted fracture at current density of 10 mA cm^{-2}. Fractographical examinations showed that increase in hydrogenation current density cause an increase in brittle character on the fracture surface. Complete brittle fractures were observed after SSRT at10 and 20 mA cm^{-2}.

1. Introduction

Duplex stainless steels (DSSs) offer an attractive combination of properties, particularly high mechanical strength, toughness and superior corrosion resistance, which have result in exploitation by the many ways of industries. These superior properties arise from an equilibrium of the austenitic (γ) and ferritic (α) phases with a 50:50 α/γ ratio. Despite good corrosion resistance, environmentally assisted cracking in the DSS by absorbed hydrogen atoms should be considered.

Hydrogen degradation is one of the causes of premature failure of steel constructions and generates large losses in many industrial areas such as: petrochemical, oil and gas production and marine. Unexpected failures have also been observed in cathodic protected constructions used inshore and offshore. Hydrogen charging occurring as a part of cathodic protection has been shown to be enough to produce hydrogen embrittlement (HE) of steels [1,2]. Considering the fact that cathodic protection is often an integral part of exploitation of DSS components in marine environments [3,4], the assessment of hydrogen assisted cracking of DSS is absolutely essential. The effect of cathodic charging on the mechanical properties of DSS have been investigated [5]. In general, cathodic charging during tensile test always leads to reduction in ductility, and cause the embrittlement of DSS to occur. The tendency to HE is time dependent and greatly influenced by strain rate [6]. Under these conditions, the ferrite phase becomes more brittle than austenite phase, particularly when hydrogen pick-up becomes significant. The present work has been aimed at further claryfing the effect of hydrogenation on the mechanical properties and cracking behavior of 2205 DSS in synthetic sea water.

2. Experimental procedure

2.1. Material

The material used in this investigation was 2205 DSS (plate of 15 mm thickness) with its chemical composition shown in Table 1. The material was solution annealed at 1100 °C for 0.5 h then water quenched to homogenize the microstructure. The respective microstructures in the transverse and the longitudinal orientations after heat treatment are shown in Fig. 1(a) and (b), respectively. As can be seen in these micrographs, the austenite phase (bright area) was embedded in the semi-continuous ferrite matrix (grey area). The volume fractions of ferrite and austenite, determined by image analyzer, were 52% and 48%, respectively.

Table 1. Chemical composition of 2205 duplex stainless steel.

Content of chemical elements (wt. %)									
C	Cr	Ni	Mo	Mn	Si	P	S	N	Fe
0.021	22.70	5.95	3.20	0.88	0.39	0.020	0,001	0.16	balance

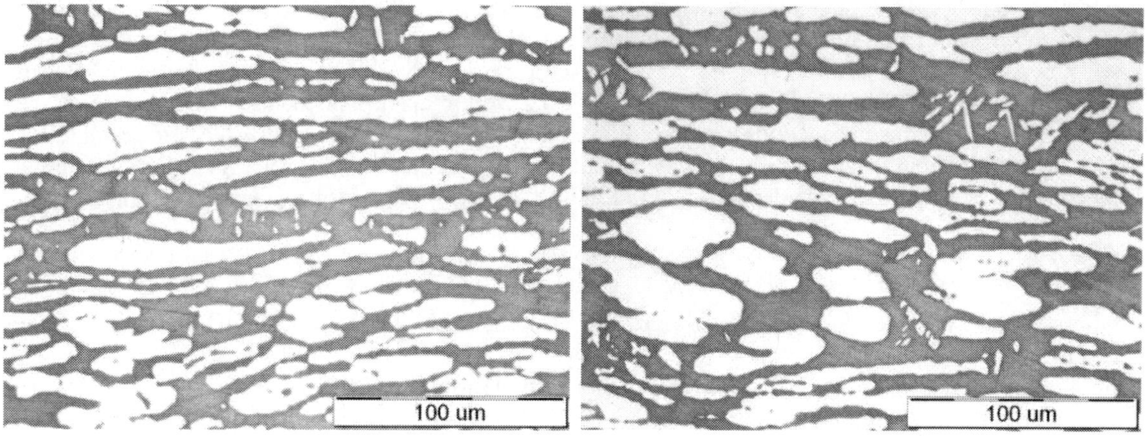

Figure 1. Microstructure of 2205 DSS: (a) transverse, (b) longitudinal orientation.

2.2. Slow strain rate tests

Slow strain rate testing (SSRT) technique according to EN ISO 7539-7 was employed in aim to evaluate the susceptibility to hydrogen assisted stress corrosion cracking of the steel investigated. Cylindrical specimens (Fig. 2) of 10.0 ± 0.2 mm outer diameter with a gauge section of 5.0 ± 0.2 mm and 50.0 ± 0.2 mm length, were used for the SSRT method.

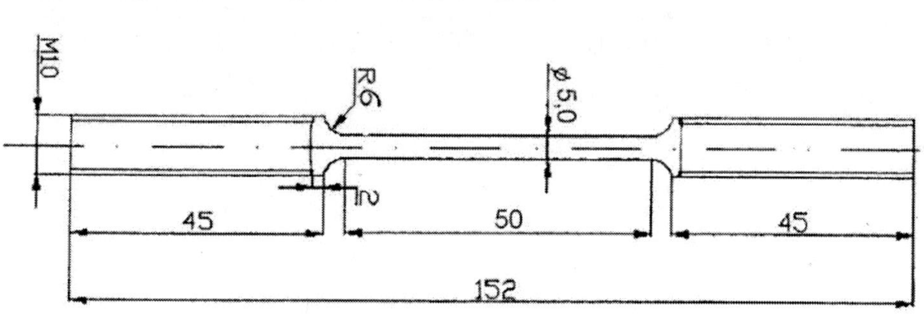

Figure 2. Specimen for SSR tests.

SSRTs were conducted in air and under various cathodic current densities conditions (from 0.1 to 20 mA cm^{-2}), at a constant strain rate of $1*10^{-6}$ s^{-1}. The solution for SSRT was synthetic sea water (type A acc. to PN-66/C-06502) with chemical composition given in Table 2.

Table 2. Chemical composition of synthetic sea water (type A).

Components	Concentration, g dm^{-3}
NaCl	24.53
$MgCl_2$	5.20
Na_2SO_4	4.00
$CaCl_2$	1.16
KCl	0.695
$NaHCO_3$	0.201
KBr	0.101
H_3BO_3	0.027
$SrCl_2$	0.025
NaF	0.003

After a final rinse with acetone, the tensile specimen was immersed in test environment in a specially designed chamber mounted in a tensile testing machine grips. SSRT tests were continued until the specimens failed. Tests in each of the cathodic current density condition were duplicated. Ultimate tensile strength (UTS) and uniform elongation (UEL) were recorded in each test. Reduction of area at fracture (RA) and the energy consumed to failure the specimen (E) were calculated after each test.
The hydrogen embrittlement index was calculated by the percentage change of the determined mechanical parameter after hydrogenation in relation to that obtained for uncharged condition:

$$HEI = [(X-X_H)/X)]\cdot 100 \quad (1)$$

where X and X_H are the determined mechanical parameter in air and after hydrogen charging, respectively.
Fractographic examinations of specimens after completion of SSRT tests were carried out using Hitachi S-3400Nscanning electron microscopy (SEM) in order to characterize the mode of rupture.

3. Results and discussion

The results of SSRT illustrating the effect of hydrogen charging conditions of the changes of the ultimate tensile strength (UTS), the ultimate elongation (UEL), the reduction of area (RA) and the energy to failure (E) values are summarized in Table 3.

Table 3. Effect of applied cathodic current on the variations of UTS, UEL and RA after SSRT in synthetic sea water.

Cathodic current density (mA cm^{-2})	UTS (MPa)	UEL (%)	RA (%)	E (J)
Air	725	26.3	51.4	166.9
0.1	710	22.3	31.1	137.6
1.0	704	12.8	25.2	75.6
5.0	691	16.7	20.8	96.9
10.0	686	14.1	18.0	80.6
20.0	714	16.9	13.5	102.0

The stress vs. percentage of elongation curves of DSS obtained in air and in synthetic sea water are shown in figure 3. The SSRT results indicated that the tensile behavior of 2205 DSS was a strong function of the applied cathodic current density. At 0.1 mA cm^{-2}, the value of ultimate tensile strength (UTS) was 710 MPa, very close to those obtained from the air test.
It was proved that hydrogen charging conditions did not influence the UTS values, despite minor fluctuations. By the contrast, detrimental effect of hydrogen was distinctly observed in case of UEL and RA values. As shown in previous results [7], RA proved to be the most sensitive measure of embrittlement as caused by the uptake of hydrogen by the steel. The susceptibility of 2205 DSS to hydrogen assisted cracking with applied hydrogenation current in synthetic sea water, as manifested by the change of reduction of area (RA) and ultimate elongation (UEL) values after SSRT tests, are shown in figure 4.

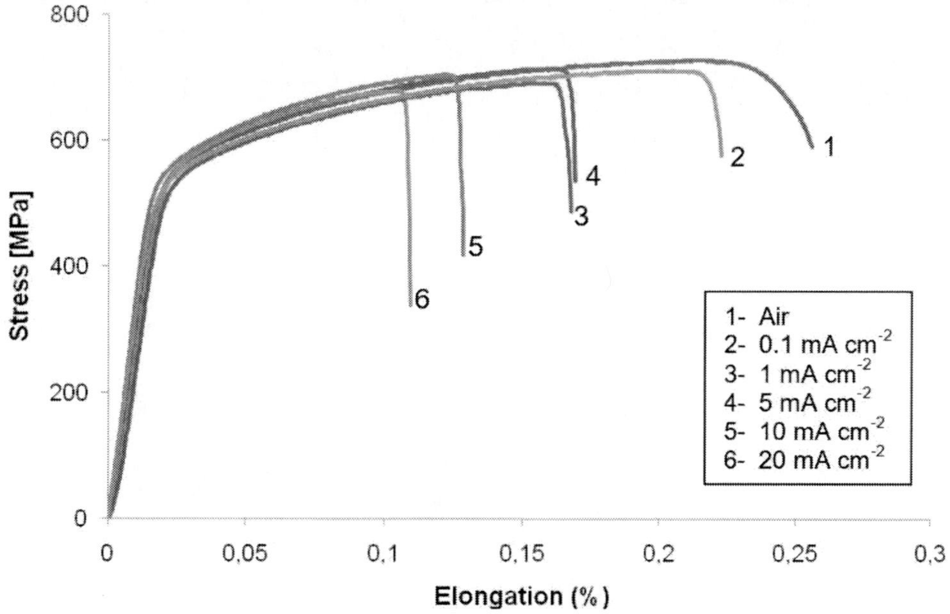

Figure 3. SSRT stress-elongation curves for 2205 DSS in synthetic seawater at various applied hydrogenation currents.

Constant decrease in reduction of area was observed with an increase in hydrogenation current density, while ultimate elongation values were not so stable. The first drop in UEL was observed after SSRT performed at 1 mA cm^{-2}, then the value of UEL raised again with increasing current density. Therefore, to compare effect of hydrogen charging, the hydrogen embrittlement index (HEI) was calculated using the reduction of area (RA) values (Fig. 5). As seen in figure 5, hydrogen embrittlement index increased with an increase in cathodic current density.

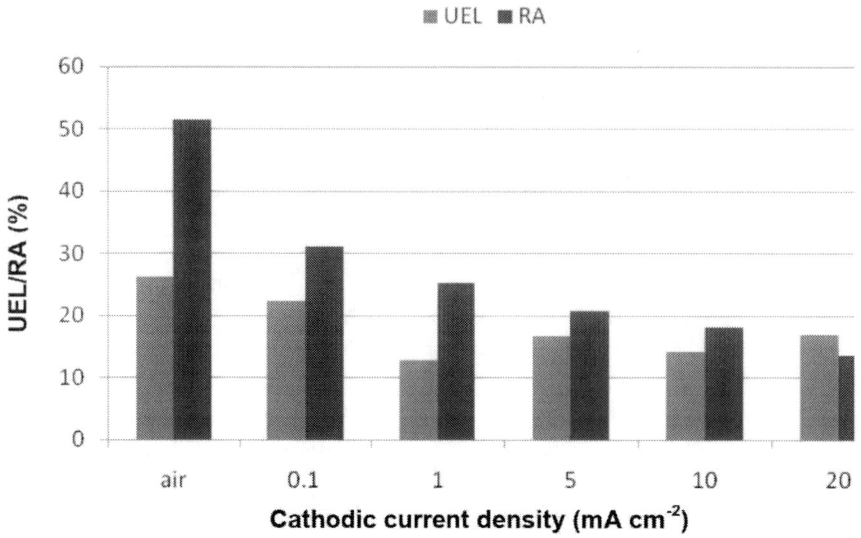

Figure 4. Effect of applied hydrogenation current on the changes of UEL and RA values of 2205 DSS after SSRT in synthetic sea water.

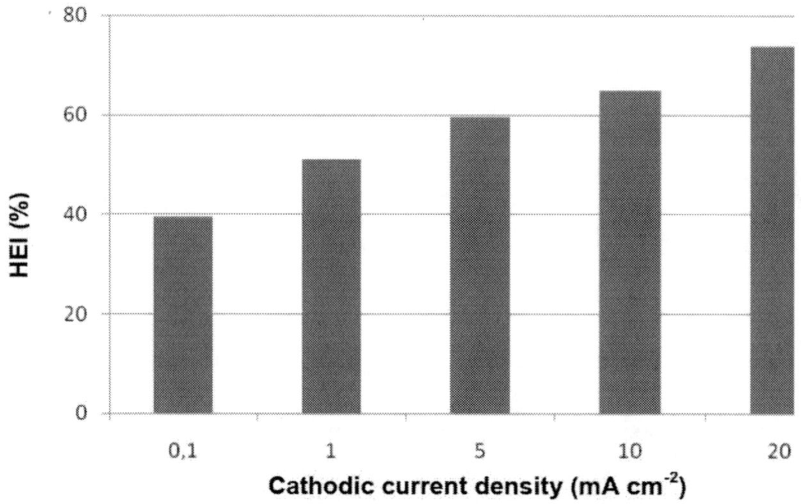

Figure 5. Effect of applied hydrogenation current on the HEI values of 2205 DSS after SSRT in synthetic sea water.

The results of mechanical test were in a good agreement with the results of the fractographic observation. SEM fractographs of the fracture specimen of 2205 DSS after SSRT in synthetic sea water are shown in figure 6. Distinct influence of hydrogen on the fracture morphology was observed. In case of the as-received material after SSRT in air, the ductile fracture mechanism was dominant as shown in Figure 6a, where a large number of dimples can be observed. The fracture surface of hydrogen charged specimen (Fig. 6b-d) have changed. The increase in hydrogenation current caused an increase in brittle character of fractures observed. The fracture after SSRT at 20 mA cm^{-2} showed predominant cleavage areas, suggesting that the brittle fracture mechanism had become dominant. Fractures after hydrogen charging revealed also a large number of secondary cracks.

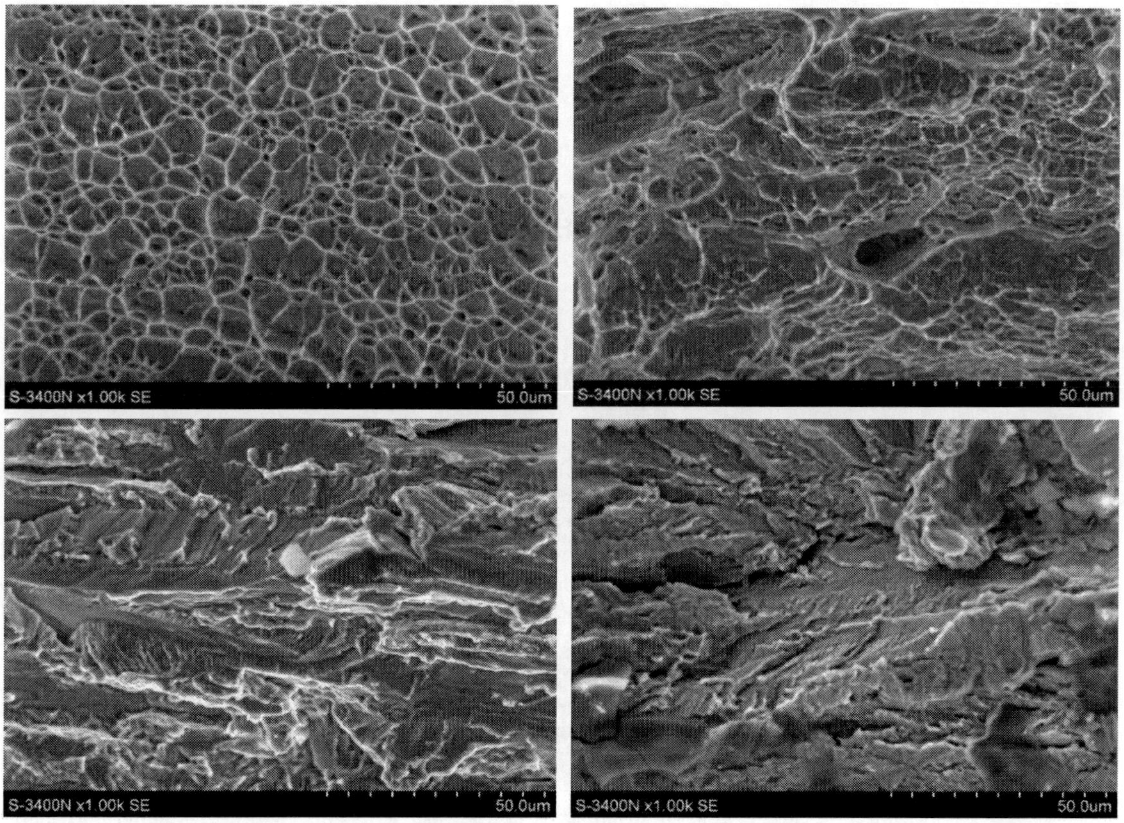

Figure 6. Fractograph after SSRT of 2205 DSS: (a) uncharged-air and hydrogen charged at: (b) 5 mA cm^{-2}, (b) 10 mA cm^{-2}, (d) 20 mA cm^{-2}.

4. Conclusions

The susceptibility to hydrogen assisted cracking of 2205 DSS in synthetic sea water was strongly dependent on the cathodic current density. The higher cathodic current density the lower ductility and the more susceptibility to hydrogen assisted cracking of the steel investigated. Detrimental effect of hydrogen was distinctly observed in case of UEL and RA values, while UTS values remained almost unchanged. Hydrogen changed also the mechanism of microcracking in 2205 DSS. Ductile failure as revealed by significant extent of necking and dimples was obtained in the air test. However, at the cathodic current densities, brittle fracture mechanism had become dominant, what was highlighted with an increase in cathodic current density applied during SSRT.

Acknowledgements

Scientific work was supported by the Polish Ministry of Science and Higher Education in the years 2010-2012 under the research project No. IP 2010 025870.

References:

[1] Domżalicki P, Łunarska E and Birn J 2007 *Mater. Corr.* **58** 413.
[2] Zucchi F, Grassi V, Monticelli C and Trabanelli G (2006) *Corr. Sci.* **48** 522.
[3] Turnbull A, Griffiths A and Reid T 1999, NACE Corrosion '99, paper No. 148.
[4] Billingham J, Sharp J V, Spurrier J and Kilgallon P J 2003 report No. 105, 2003, Marine Technology Centre, Cranfield University, UK.
[5] Chou S-L and Tsai W-T 1999 *Mater. Chem. Phys.* **60** 137.
[6] El-Yazgi A A and Hardie D 1996 *Corr. Sci.* **38** 735.
[7] Michalska J 2006 The assesment of hydrogen embrittlement resistance of duplex stainless steel, PhD thesis, Silesian University of Technology, Katowice, Poland, 2006.

IOP Conf. Series: Materials Science and Engineering **35** (2012) 012014 doi:10.1088/1757-899X/35/1/012014

Low-cycle fatigue life of steel P92 under gradual loading

M Cieśla, G Junak
Department of Materials Technology, Silesian University of Technology, Krasinskiego 8, 40-019 Katowice, Poland

E-mail: marek.ciesla@polsl.pl

Abstract. The aim of the study was to determine low-cycle fatigue life characteristics of P92 steel used in the power unit components that work under the highest effort. Steel service life was determined in the tests for total fixed strain $\Delta\varepsilon_t$ ranges from 0.6 to 1.2 % and with application of two-stage sequence loading. Low-cycle fatigue tests were conducted at room temperature and at 550 °C on MTS-810 testing machine. The tests under sequence loading were carried out for two strain ranges: $\Delta\varepsilon_t$ = 0.6 and 1.2 %. Sinusoidal load cycles of a coefficient R = -1 were used. Low-cycle fatigue life characteristics were developed with consideration of loading sequence impact. It was found out the fatigue life of an element under sequence loading is strictly correlated with its history of strain, which determines the degree of material microstructure damage. Higher degree of steel damage was found in two-stage tests, in which the larger range of strain $\Delta\varepsilon_t$ = 1.2 % was used as the first one. Fractography analyses of scrap revealed the patterns of fatigue lines and bands typical for this process as well as numerous secondary cracks occurred to the boundaries of former austenite grains. Analytic method of predicting low-cycle fatigue life of an element under two-stage sequence loading is presented. For this purpose, an energy criterion for the material life developed for tests carried out in conditions of low-cycle fatigue was used.

1. Introduction

Destruction of industrial objects after long-term operation under variable mechanical and/or thermal loads is the issue bothering both, the industrial plants and scientific centres. This issue is of particular importance for power generation sector where replacement of a given element entails huge costs.[1÷5]. There have been numerous attempts to describe predicting the life of industrial objects. Yet, the issue of fatigue life considering the load history remains unresolved due to its complexity and remains a subject-matter of researches and analyses. Therefore, the description of fatigue phenomenon, which, basing upon specific calculations, demonstrates the impact of load history on degree of material damage which determines the life of a given element seems advisable [6÷13]. In this case, fatigue characteristics of materials for the parameters, which simulate operating conditions of the facility being analysed have to be developed [14÷16].

The currently applicable world standards and procedures regarding the assessment of the life of industrial facilities capture this phenomenon in many different ways. The German and British standards (TRD and BS 7910, respectively) should be mentioned herein. These procedures address the problem of fatigue crack nucleation and further stages of fissure development.

In this paper, the low-cycle fatigue life characteristics of new steel grade P92, which, due to its good strength at elevated temperatures, is used in power plants for completion of pipelines and elements of conventional power boilers are presented. Fatigue tests were carried out for constant strain ranges $\Delta\varepsilon_t$ and under two-stage sequence loading. The results obtained may be used for predicting the life of power engineering equipment made of this steel grade.

2. Analysis of low-cycle fatigue process of P92 steel

Cylindrical samples (figure 1) made of steel in grade P92 in as toughening conditions were used in the tests. Microstructure of the steel is shown on figure 2, and its chemistry and basic mechanical properties are given in tables 1 and 2.

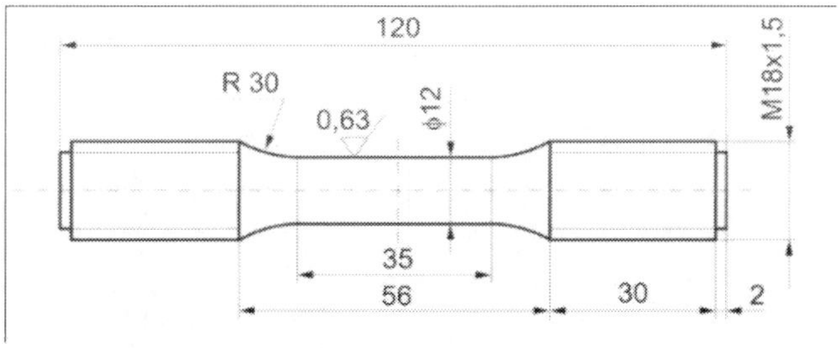

Figure1. Sample for low-cycle fatigue tests.

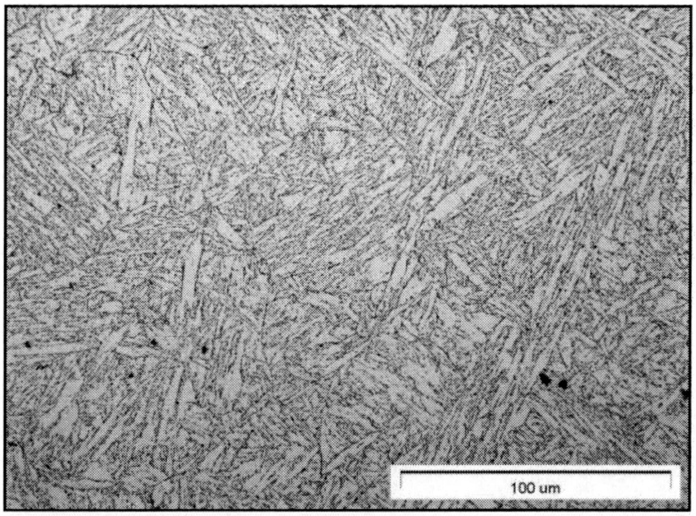

Figure 2. Microstructure of P92 steel - tempered martensite.

Table 1. Chemical composition of steel P92.

Contents of elements (wt. %)								
C	Si	Mn	P	S	Ni	Cr	Mo	W
0.11	0.30	0.60	0.017	0.004	0.20	9.50	0.50	1.90

Table 2. Basic mechanical properties of steel P92.

Mechanical properties			
R_m (MPa)	$R_{0.2}$ (MPa)	A_5 (%)	Z (%)
at room temperature			
705.7	529.7	25.4	67.5
at 550 °C			
422.0	356.0	41.7	76.5

Fatigue tests at room temperature and at 550 °C were performed in tensile/compression conditions with asymmetry factor of R = -1. The tests were conducted in testing machine under controlled strain with use of an extensometer of a gauge length of 25 mm. Monotone tests were carried out for 5 levels of total strain from the range $\Delta\varepsilon_t = 0.6 \div 1.2$ %. Low-cycle fatigue life characteristics N_f for steel were developed basing upon the results obtained and summarized on figure 3.

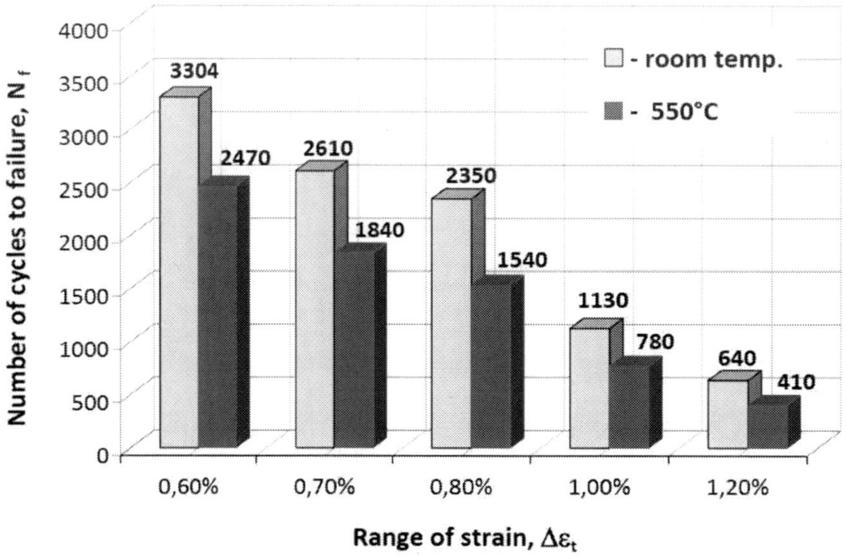

Figure 3. Low-cycle fatigue life of steel P92 at room temperature and at 550 °C.

Steel of P92 in conditions of monotone tests at both, the room temperature and at 550 °C dropped cyclically. Mathematical models of cyclic strain described with the formula $\sigma_{an} = K'\varepsilon_p^{n'}$ are given in table 3.

Table 3. Characteristics of cyclic strain of P92 steel.

$\sigma_{an} = K'\varepsilon_p^{n'}$	
σ_{an} – saturation stress (MPa)	
K' - cyclic strength coefficient (MPa)	n' - cyclic weakening coefficient
room temperature	
953.5	0.11
temperature of 550 °C	
1051.9	0.20

In the next, principal stage of the study, low-cycle fatigue tests under sequence loading were conducted at room temperature and at the temperature of 550 °C. Two ranges of total strain were used $\Delta\varepsilon_t = 0.6$ % and 1.2 %.

In variant **A (L-H)** (figure 4a and 5a), the fatigue tests were conducted at first for a strain range of $\Delta\varepsilon_t = 0{,}6$ % with the assumed number of cycles N_1, and next, the tests were continued for the strain range of $\Delta\varepsilon_t = 1.2$ % until failure of the sample after N_2 cycles.

In variant **B (H-L)** (figure 4b and 5b) – reverse direction of application of total strain $\Delta\varepsilon_t$ was used.

For each variant of the research, in its first stage, the fatigue test was carried out for a given strain range $\Delta\varepsilon_t$ (0.6 or 1.2 %) with the assumed number of cycles N_1. For each variant (L-H or H-L), 1st stage of low-cycle fatigue test was carried out with various assumed numbers of cycle N_1. It was the basis for determination of degree of material damage $\frac{N_1}{N_{1f}}$. Subsequently, the fatigue test was continued in the 2nd stage of the given variant until failure of the sample after N_2 number of cycles. This enabled calculation of the material damage degree $\frac{N_2}{N_{2f}}$. In the ratios of $\frac{N_i}{N_{if}}$ the values of N_{1f} and N_{2f} mean the number of cycles in conditions of single-stage fatigue test conducted for a given total strain range $\Delta\varepsilon_t$ after which the sample failure occurred.

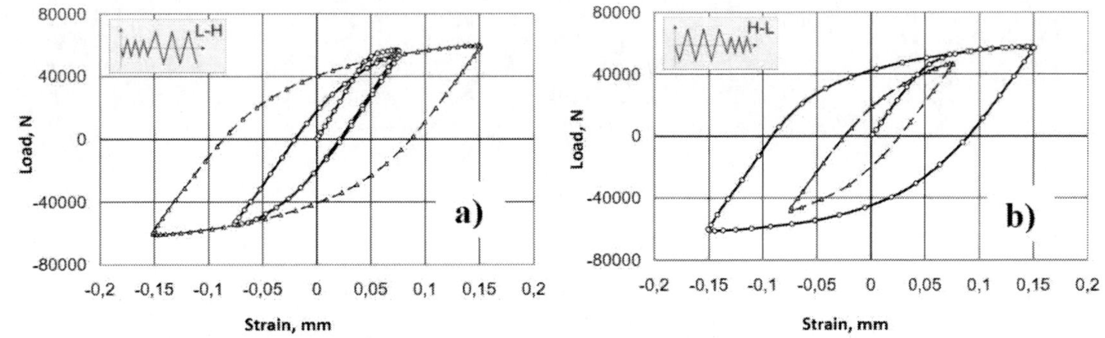

Figure 4. Hysteresis loop recorded in conditions of two-stage low-cycle fatigue tests of P92 steel at room temperature.

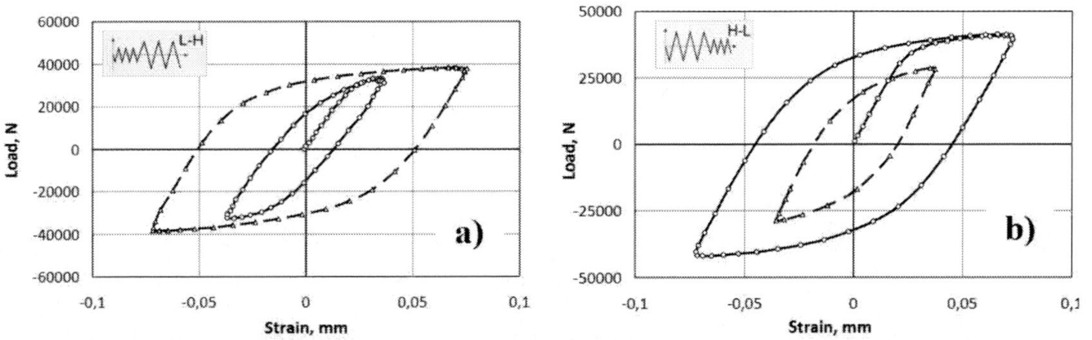

Figure 5. Hysteresis loop recorded in conditions of two-stage low-cycle fatigue tests of P92 steel at the temperature of 550 °C.

The obtained results constituted the basis for developing graphs of low-cycle fatigue life of P92 steel at sequence loading (figures 6 and 7). Developed fatigue life characteristics were plotted against the graph consistent with the Palmgren-Miner hypothesis [9,10].

Analysis of the graphs obtained indicates that at both, room temperature and at the temperature of 550°C the graph representing low-cycle fatigue life determined in accordance with the procedure **A** (L-H) lies above the straight line illustrating the fatigue life determined pursuant to the Palmgren-Miner hypothesis, while the graph illustrating fatigue life determined pursuant to the procedure **B** (H-L) lies below this straight line. It unambiguously indicates the degree of material damage $\frac{N_1}{N_{1f}}+\frac{N_2}{N_{2f}}\neq 1$ in fatigue test is different and depends on the sequence of load applied to the sample.

The damage accumulation law of Palmgren-Miner $\frac{N_1}{N_{1f}}+\frac{N_2}{N_{2f}}=1$ does not apply to this case [17÷19]. The history of strain has a considerable impact on fatigue life of steel P92. Material damage expressed by the relation $\frac{N_1}{N_{1f}}+\frac{N_2}{N_{2f}}<1$ is larger for strain sequence "H-L" than for the reverse sequence of "L-H" $\frac{N_1}{N_{1f}}+\frac{N_2}{N_{2f}}>1$ (figures 6 and 7).

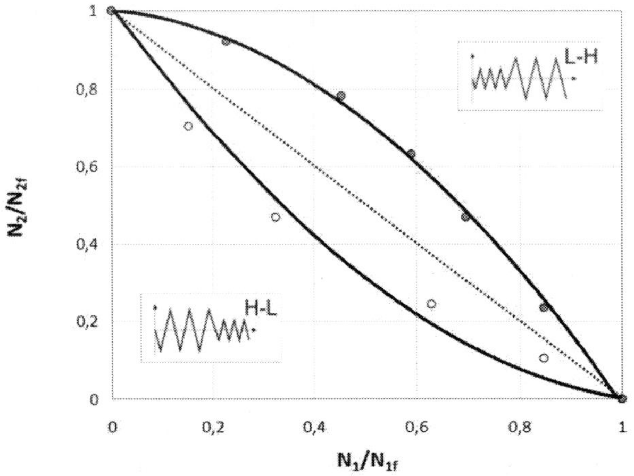

Figure 6. Characteristics of low-cycle fatigue life of P92 steel under sequence loading at room temperature.

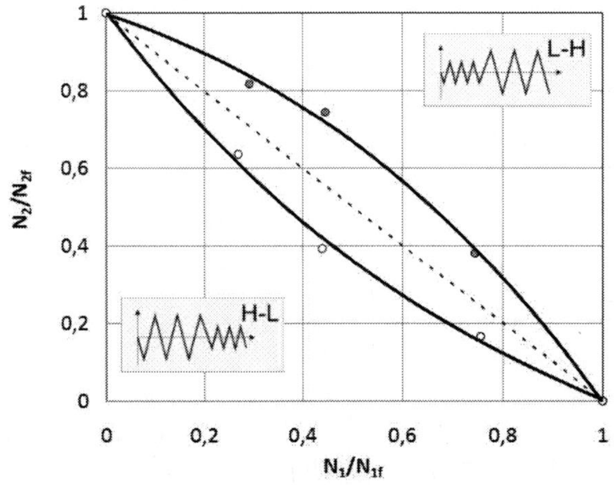

Figure 7. Graph of low-cycle fatigue life of P92 steel under sequence loading at the temperature of 550 °C.

Impact of loading sequence on material conditions, which determines its low-cycle fatigue life N_f is reflected in cyclic strain graphs. Figures 8 and 9 illustrate examples of selected cyclic strain characteristics of P92 steel under monotone and two-stage loading at room temperature and at the temperature of 550 °C. These graphs indicate that the process of cyclic strain in the 2nd stage of the fatigue test under "H-L" loading sequence runs with a lower stress value σ than in case of monotone-single stage loading. Material weakening observed in this case may be caused by the change of steel dislocation structure (figure 8) and its disruption caused by nucleation of micro-cracks (figures 9, 12 and 13). At the temperature of 550 °C material weakening under „H-L" loading sequence may result also from dynamic recovery.

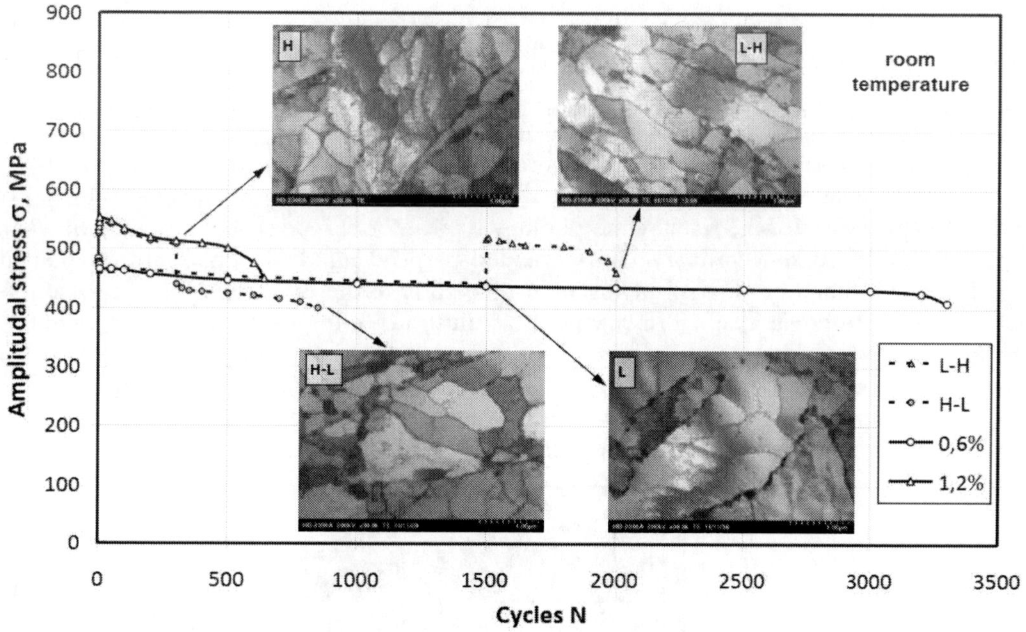

Figure 8. Characteristics of cyclic strain of P92 steel under monotone and two-stage (sequence) load at room temperature. The graph illustrates cellular dislocation structure of steel for the given variant of the test.

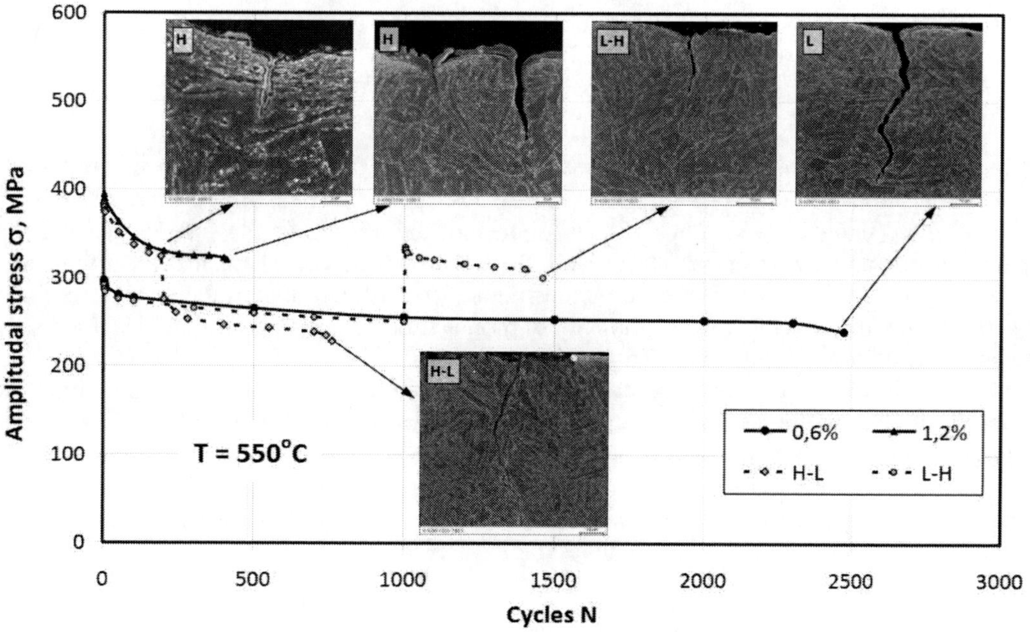

Figure 9. Characteristics of cyclic strain of P92 steel under monotone and two-stage load at the temperature of 550 °C. The graph illustrates the steel microstructure in sub-surface area for the given variant of the test.

In conditions of "H-L" two-stage test, the sub-grain groups of dislocation structure separated by low-angle boundaries were found inside extended grains after cyclic strain in „H" stage i.e. for $\Delta\varepsilon_t$ = 1.2 % (figure 8). Fine carbide grains were found at grain and sub-grain boundaries. Further run of fatigue process in "L" stage i.e. for $\Delta\varepsilon_t$ = 0.6 % led to the formation of sub-structure, in which the sub-grain boundaries were thin, and the sub-grains had smaller dislocation density.

In the tests conducted under sequence loading, the change of load sequence caused the change of material conditions, and consequently, the change of crack growth rate. It resulted in the change of exponent and constant values in the formulas describing crack propagation e.g. in the Paris formula: $\frac{dl}{dN} = C(\Delta K)^m$. This factor (crack growth rate) determined, inter alia, differences in low-cycle fatigue life N_f of P92 steel observed in test conditions of "L-H" and "H-L" (figures 6 and 7).

Fractographic analysis of sample fractures after low-cycle fatigue tests under sequence loading at room temperature revealed the fracture morphology typical for fatigue cracking of material. Numerous systems of parallel running primary faults oriented perpendicularly to crack propagation direction (figure 11a) were found on scraps running perpendicularly to the direction of the largest elongations (figure 10) in macroscopic image. Primary faults as illustrated on figures 11a and 11c were formed on the boundaries of former austenite grains. Fatigue band systems with inter-band distance of approx. 1 μm were found between individual faults i.e. in area of stable fatigue crack growth (figures 11b,c,d). Numerous secondary cracks (figures 11b,d) were found at scrap surface on the boundaries of former austenite grains, which have lower cohesion strength than the grains themselves.

Figure10. Fatigue fault of sample submitted to low-cycle fatigue test.

Observations of microstructure of samples submitted to the fatigue test at the temperature of 550°C confirmed the formation of micro-cracks just below the sample surface in first stage of "H-L" test i.e. for $\Delta\varepsilon_t$ = 1.2 % (figures 12a,b). Thus, in the 2nd stage of the test i.e. for $\Delta\varepsilon_t$ = 0.6 % accelerated crack propagation was already observed. As a result, the degree of material damage was in this case larger than in case of "L-H" variant. The observations made demonstrated that at the temperature of 550°C, likewise at the room temperature, the crack propagation in certain growth stages follows the boundaries of former austenite grains (figure 13).

On the basis of the results obtained, the analytic method of predicting (estimating) low-cycle fatigue life of P92 steel under two-stage loading was presented. In the analyses, the characteristics of low-cycle fatigue determined in monotone (single-stage) tests conducted for strain range of $\Delta\varepsilon_t$ = 0.6 and 1.2 % were taken into account. The procedure of predicting a low-cycle fatigue life of steel based on energy criterion described in [15] was used for calculations. Pursuant to the foregoing procedure, the amount of energy accumulated in the sample after the time of N_f cycles is described by the following equation:

$$W_k = \int_1^{N_f} P(\Delta\varepsilon_t, N)\cdot dN = const. \quad (1)$$

where: N_f – number of cycles to sample failure for the given total strain range $\Delta\varepsilon_t$.

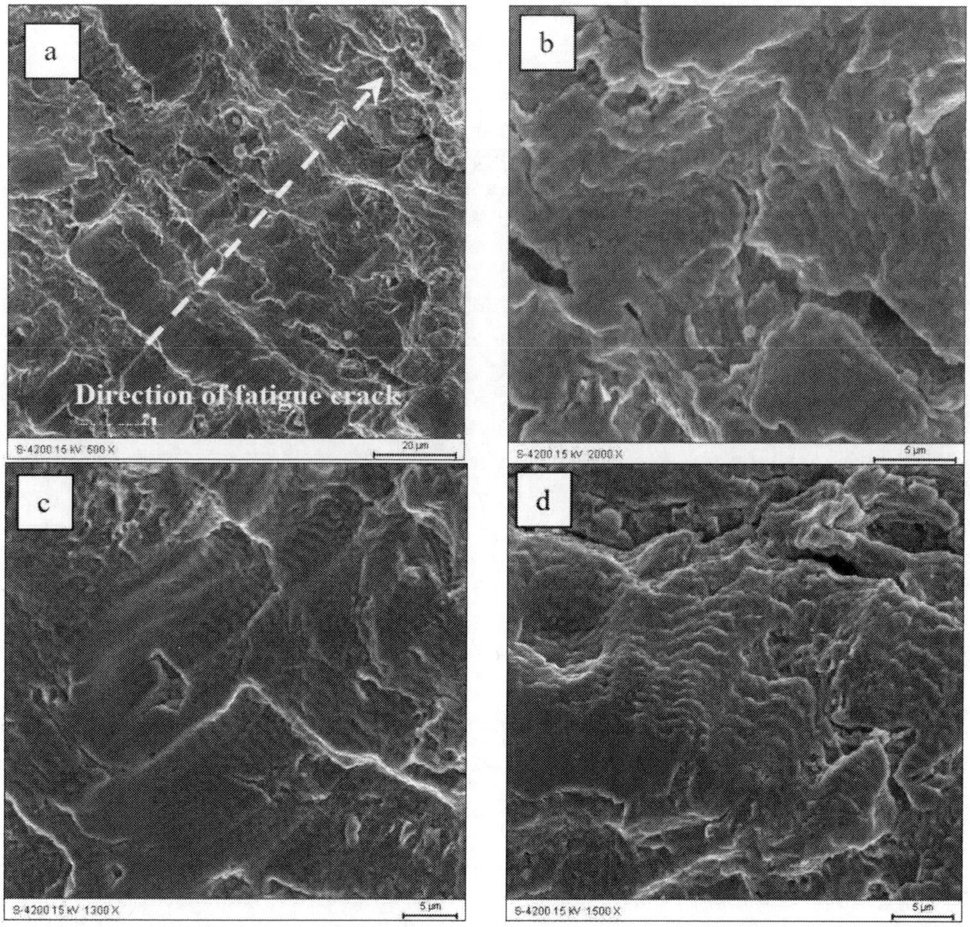

Figure 11. Microstructure of P92 steel fatigue fault at room temperature produced in B "H-L" loading sequence. (a) - systems of parallel running primary faults, (b,c,d) - fatigue band patterns in area of former austenite grains, secondary cracks.

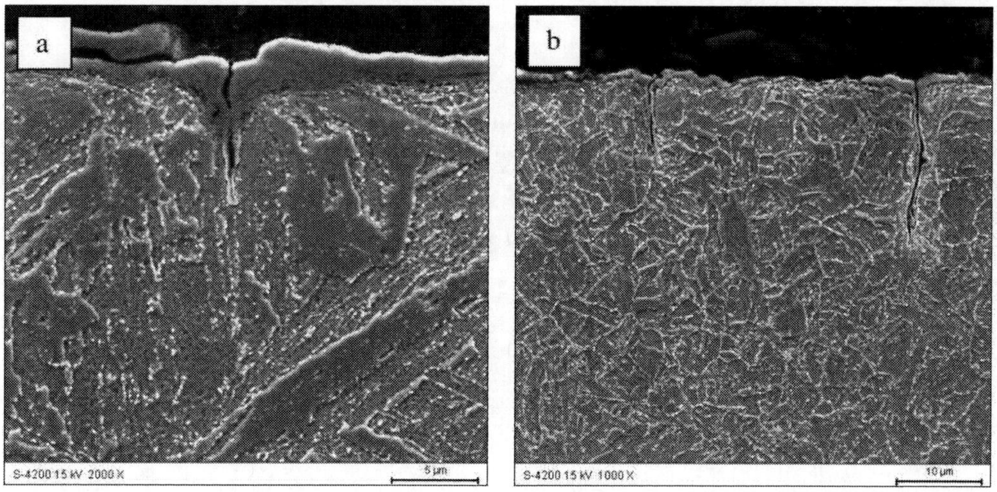

Figure 12. Microstructure of P92 steel after low-cycle fatigue tests at 550°C in "H-L" variant. Sub-surface micro-cracks formed in 1st stage of the test for $\Delta\varepsilon_t = 1.2$ %.

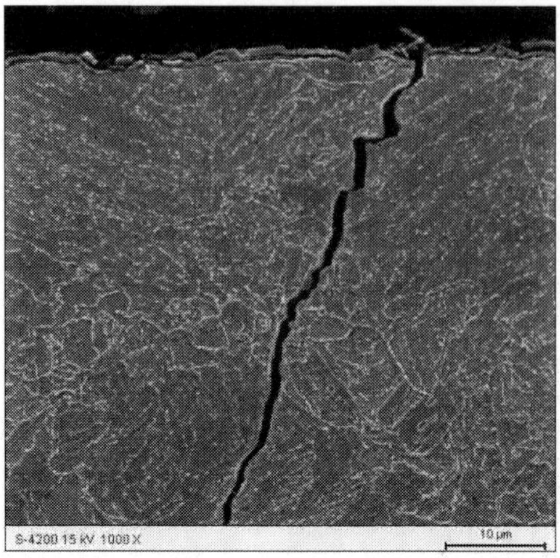

Figure 13. Propagation of fatigue cracks in P92 steel after low-cycle fatigue tests at 550 °C in "H-L" variant.

The amount of damage energy accumulated in the sample in conditions of low-cycle fatigue leading to the sample failure may be calculated in accordance with the following formula:

$$W_k = c \cdot \Delta\varepsilon_t^{\ d} \cdot N_f + a \cdot e^{N_f \cdot b} \cdot \frac{1}{b} - c \cdot \Delta\varepsilon_t^{\ d} - a \cdot e^{b} \cdot \frac{1}{b} \tag{2}$$

which is obtained by integrating the energy density function P like in formula (1) in the following form:

$$P = c \cdot \Delta\varepsilon_t^{\ d} + a \cdot e^{N \cdot b} \tag{3}$$

Constant values: a, b, c and d were obtained from the system of 4 equations, to which the input data were determined on the basis of single-stage low-cycle fatigue tests conducted for 4 ranges of total strains $\Delta\varepsilon_t$. Simultaneously, the assumption was made that the amount of energy W_k is equal to the amount of accumulated damage energy W_k leading to sample failure determined in static tensile test [19]:

$$W_k = 0{,}5 \frac{\sigma_{zer}^{\ 2}}{E} \tag{4}$$

In calculations for two-stage variant of low-cycle fatigue of test (conducted for strain ranges $\Delta\varepsilon_t = 0.6$ and 1.2 %) the amount of energy accumulated in steel W_1 was determined.

The amount of energy W_1 accumulated in steel after number of cycles $N_{1(\Delta\varepsilon_1)}$ in 1st stage of the low-cycle fatigue test conducted for the range of strain $\Delta\varepsilon_{t1}$ was calculated pursuant to the following formula (2):

$$c \cdot \Delta\varepsilon_1^{\ d} \cdot N_1 + a \cdot e^{(N_1 \cdot b)} \cdot \frac{1}{b} - c \cdot \Delta\varepsilon_1^{\ d} - a \cdot e^{b} \cdot \frac{1}{b} = W_{1(\Delta\varepsilon_1)} \tag{5}$$

Subsequently, the number of cycles $N_{1(\Delta\varepsilon_2)}$ corresponding to the same amount of energy (W_1) but for the other range of total strain $\Delta\varepsilon_{t2}$ used in the tests was calculated according to the formula (5). For calculations, the assumption was made that $W_{1(\Delta\varepsilon_1)} = W_{1(\Delta\varepsilon_2)}$.

The results of calculations for 1st stage of analysed test variant constituted the basis for computing the number of cycles leading to sample failure $N_{2(\Delta\varepsilon_2)}$ in the 2nd stage of this test variant according to the following formula:

$$N_{2(\Delta\varepsilon_2)} = N_{2f(\Delta\varepsilon_2)} - N_{1(\Delta\varepsilon_1)} \tag{6}$$

The number of cycles to sample failure in conditions of single-stage test for the formula (6) $N_{2f(\Delta\varepsilon_2)}$ was computed with use of the following formula:

$$c \cdot \Delta\varepsilon_2^{\ d} \cdot N_{2f(\Delta\varepsilon_2)} + a \cdot e^{(N_{2f(\Delta\varepsilon_2)} \cdot b)} \cdot \frac{1}{b} - c \cdot \Delta\varepsilon_2^{\ d} \cdot N_{1(\Delta\varepsilon_2)} - a \cdot e^{(N_{1(\Delta\varepsilon_2)} \cdot b)} \cdot \frac{1}{b} = W_2 \tag{7}$$

where: $W_2 = W_K - W_1$ – the amount of energy accumulated in steel in 2nd stage of tests conducted for strain range of $\Delta\varepsilon_{t2}$.

Basing upon the results obtained, the coordinates (N_2/N_{2f}) of P and R points written down to the fatigue life characteristic (figure 14) developed on the basis of experiments (figure 6) were determined. Good consistency of (N_2/N_{2f}) ratio with its value determined on the basis of the results obtained in 2nd stage of fatigue test was confirmed for both load sequences "H-L" (point R) and "L-H" (point P).

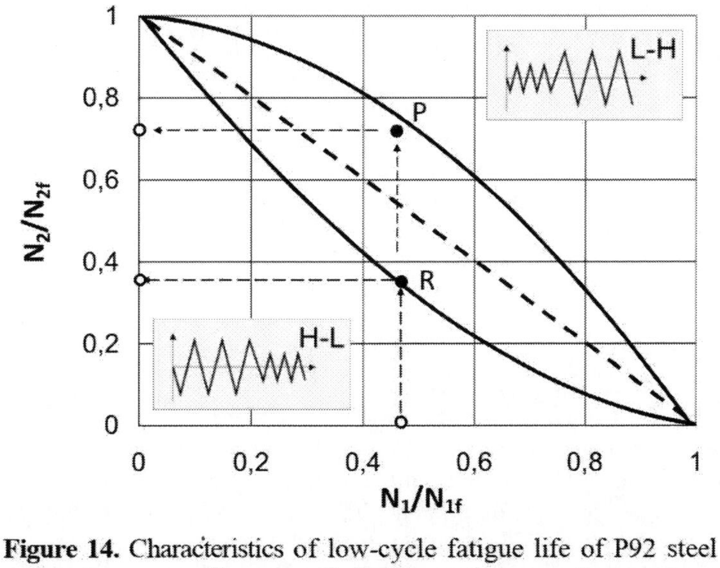

Figure 14. Characteristics of low-cycle fatigue life of P92 steel under sequence (two-stage) loading at room temperature. Coordinates of P and R points are consistent with analytical prediction of fatigue life in accordance with the energy hypothesis of fatigue damage accumulation.

3. Conclusions

- P92 steel in conditions of low-cycle fatigue tests conducted for fixed and sequentially changed ranges of total strain $\Delta\varepsilon_t$ shows cyclical weakening at both, the room temperature and at 550 °C.
- The degree of material damage found in fatigue test of P92 steel was different and dependent on the sequence of load applied to the sample $\frac{N_1}{N_{1f}} + \frac{N_2}{N_{2f}} \neq 1$. Palmgren-Miner's law of damage accumulation does not apply to this case. The history of strain has a considerable

impact on low-cycle fatigue life of steel P92. Material damage expressed by the relation $\frac{N_1}{N_{1f}}+\frac{N_2}{N_{2f}}<1$ is larger for "H-L" strain sequence than for the reverse "L-H" sequence $\frac{N_1}{N_{1f}}+\frac{N_2}{N_{2f}}>1$.

- Material weakening in the 2nd stage of the tests conducted in "H-L" variant was larger than it was found-out in case of single-stage tests carried out in "L" variant. This results in shorter low-cycle fatigue life of steel N_f obtained in conditions of "H-L" tests than the fatigue life obtained in conditions of a single-stage fatigue test in "L" variant.
- The steel weakening observed in "H-L" test conditions is caused by the change of steel dislocation structure and its disruption by nucleation of micro-cracks at surface area.
- Service life of P92 steel predicted in conditions of low-cycle fatigue under sequence loading with use of energy criterion of fatigue damage accumulation presented in this paper is consistent with the results of experiments with acceptable accuracy.

References

[1] Renowicz D, Cieśla M 2007 *Int. J. of Computational Mat. Sci. and Surf. Eng. (IJCMSSE)*, **1** issue 4 424.
[2] Okrajni J, Mutwil K and Cieśla M 2007 *J. of Achiev. in Mat. and Manuf. Eng.* **22** issues 2 63.
[3] Renowicz D, Cieśla M 2007 *J. of Achiev. in Mat. and Manuf. Eng.* **21** issue 2 49.
[4] Renowicz D, Hernas A, Cieśla M and Mutwil K 2006 J. of Achiev. in Mat. and Manuf. Eng. **18** 219.
[5] Dobrzański J 2005 *J. of Mat. Proc. Techn.* 785.
[6] Kocańda S, Kocańda D and Tomaszek H 1996 *Proc. Finfth Int. Conf. on Structural Failure* Vienna, Austria.
[7] Marquis G and Solin J 1999 *Fatigue Design and Reliability*. Elsevier ESIS Publication **23**.
[8] Padmadinata.U H, Schijve J 1994 *NASA Conf. Publ.* 3274, (C.E. Harris, ed.) 547.
[9] Miner M A 1945 *J. Appl. Mech., Trans. ASME* **67** A-159.
[10] Palmgren A 1924 *Endurance of ball bearings* Z. Ver. dt. Ing. **68** 339.
[11] Willenborg J, Engle R M and Wood H A 1971 *AFFDL-TR71-1, Air Force Flight Dynamic Laboratory* Dayton 1971.
[12] Wheeler O E 1972 *ASME Trans., J. Of Basic Eng.* **94** 181.
[13] Junak G, Cieśla M. 2011 *Arch. of Mat. Sci. and Eng.* **48** 19.
[14] Ducki K, Cieśla M 2008 *J. of Achiev. in Mat. and Manuf. Eng.* **30** (1) 19.
[15] Okrajni J, Junak G 2008 *J. of Achiev. in Mat. and Manuf. Eng.* **26** Issue 2 147.
[16] Schijve J 2002 *Proc. of ECF 14 Conference: "Fracture Mechanics Beyond 2000"*, 211
[17] Miller K J, Zachariach K P 1977 *J. of Strain Anal.* **12** 4 263.
[18] Miller K J 1991 *An introductory lecture on fatigue of metals, course of Metal Fatigue* University of Shefild.
[19] Kocańda S 1985 *A fatigue cracking of metals* Ed. WNT, Warsaw.

IOP Conf. Series: Materials Science and Engineering **35** (2012) 012015 doi:10.1088/1757-899X/35/1/012015

The effect of strain hardening on resistance to electrochemical corrosion of wires for orthopaedics

J Przondziono[1], W Walke[2], E Hadasik[1] and J Szymszal[1]
[1]Department of Technology Materials, Silesian University of Technology, Krasińskiego 8, 40-019 Katowice, Poland
[2]Department of Biomaterials and Medical Devices Engineering, Silesian University of Technology, Akademicka 16, 44-100 Gliwice, Poland

E-mail: joanna.przondziono@polsl.pl

Abstract: The purpose of this research is to evaluate electrochemical corrosion resistance of wire with modified surface, made of stainless steel of Cr-Ni-Mo type, widely used in implants for orthopaedics, depending on hardening created in the process of drawing. Tests have been carried out in the environment imitating human osseous tissue. Pitting corrosion was determined on the ground of registered anodic polarisation curves by means of potentiodynamic method with application of electrochemical testing system VoltaLab® PGP 201. Wire corrosion tests were carried out in Tyrode solution on samples that were electrochemically polished as well as electrochemically polished and finally chemically passivated. Initial material for tests was wire rod made of X2CrNiMo17-12-2 steel with diameter of 5.5 mm in supersaturated condition. Wire rod was drawn up to diameter of 1.35 mm. This work shows the course of flow curve of wire made of this grade of steel and mathematical form of yield stress function. The study also presents exemplary curves showing the dependence of polarisation resistance in strain function in the drawing process of electrochemically passivated and electrochemically polished and then chemically passivated wire.

1. Introduction

Modern medicine features constant development of state-of-the-art diagnostic techniques and treatment methods. This development is connected with intensive development of material engineering that deals with, among other things, designing modern materials for medicine. The most popular materials used in medicine so far have been metallic materials. They are most frequently used as surgical instruments and medical appliances, but most of all they are widely used in implantology, replacing damaged parts of tissue or bones.

One of the ground criteria of evaluation of biomaterials used for implants is their biotolerance in contact with tissues and saline solutions. Despite the fact that stainless steel of Cr-Ni-Mo type is a material with the lowest corrosion resistance among metallic materials used for implants, wires that are made of it are widely used for implants for oral and maxillofacial surgery, vascular surgery as well as orthopaedics, where they can be used as intramedullary nails, wires for binding bone fractions, elements of external stabilisers or as implants used in spine stabilisation [1,2].

Mechanical properties necessary for proper operation of wires for orthopaedic implants depend on strain hardening that takes place in the drawing process. Strain applied in cold working also influences wire resistance to electrochemical corrosion.

A proper characteristics of material technological plasticity is of vital importance for the selection of optimum parameters of wire manufacturing process. Obtaining of structure prone to drawing process as well as product featuring required functional characteristics (among other things mechanical characteristics and resistance to corrosion), are both conditioned by them. Plastic strain is accompanied by the phenomenon of work hardening that is connected with the increase of yield stress σ_p [3,4]. Correct determination of plastic working parameters and acquisition of proper final characteristics of products are invariably connected with analysis of the course of function $\sigma_p=f(\varepsilon)$. The curves of changes of yield stress σ_p as the function of strain ε (so called strain hardening curves) enable to predict material behaviour during plastic working process. As mentioned before, strain in the process of drawing also significantly influences corrosion characteristics of wire. In the studies [5-7] it was proved that increase of strain brings about deterioration of corrosion characteristics of drawn material.

Production engineers who design technology of wire production make use of strain hardening curves in order to select drawing parameters in such a way that they can obtain wires with required mechanical properties for each application. For this purpose, they can also use so called technological curves, i.e. curves presenting the relation between tensile strength of wire as the function of strain [8]. This study presents a similar attitude in order to predict corrosion characteristics of wire depending on strain applied in drawing process.

Wire used in orthopaedics for short-term implants are made most frequently of stainless steel X2CrNiMo17-12-2. This study presents the course of strain hardening curve of wires made of this steel and mathematical form of yield stress function. Evaluation of resistance to electrochemical corrosion of wires made of X2CrNiMo17-12-2 steel, widely used for implants, was made, depending on strain in drawing process. The tests were carried out in the environment simulating human bone tissue.

An important issue in the process of functional characteristics forming of wires and wire products used in orthopaedics is the selection of physical and chemical characteristics of the surface [9-15]. They should be adapted to the characteristics of the environment of human bone tissue. On that account, the influence of surface modification on physical and chemical characteristics of wire was analysed in this study. The study presents examples of curves showing the relation between polarisation resistance as the function of strain in drawing process of wires electrochemically polished and electrochemically polished and then chemically passivated.

2. Materials and methods

Initial material for tests was wire rod made of steel X2CrNiMo17-12-2 with diameter of 5.5 mm, supersaturated. Wire rod was drawn to diameter of 1.35 mm. Total logarithmic strain in drawing process was ε=2,81. During drawing process, samples were cut off for mechanical and corrosion tests. Chemical composition of the steel is presented in Table 1.

Table 1. Chemical composition of X2CrNiMo17-7-2 steel.

Element	C	Mn	Si	P	S	Cr	Ni	Mo
X2CrNiMo17-7-2	0.02	1.57	0.44	0.03	0.001	16.57	11.02	2.12

Samples were ground by means of abrasive paper. Next, they were cleaned in ultrasound washer and electrochemically polished. Chemical composition and process parameters are presented in Table 2.

IOP Conf. Series: Materials Science and Engineering **35** (2012) 012015 doi:10.1088/1757-899X/35/1/012015

Table 2. Composition of solution and parameters of wire polishing process.

Substance	Quantity	Process parameters
Phosphoric acid	55÷60% mas.	Temperature: 60 ± 1°C,
Sulfuric acid	35÷45% mas.	Current strength:
Oxalic acid	40÷60 g/dm^3	40 ± 1 A/dm^2,
Acetanilide	40÷60 g/dm^3	Time: 4 min.
Corrosion inhibitor	3 g/dm^3	

After electrochemical polishing, samples were cleaned again in ultrasound washer. Then, their surface was passivated in nitrous acid for 60 min. Metal reacted with acid, which led to creation of oxide layer that prevented further reaction. Evaluation of resistance to electrochemical corrosion of wires drawn of steel Cr-Ni-Mo was made with application of the system for electrochemical tests VoltaLab® PGP 201 by Radiometr [16,17]. The tests were carried out in Tyrode solution simulating human bone tissue, at the temperature of 37±1°C and pH 6.9. Chemical composition of Tyrode solution was given in Table 3.

Table 3. Chemical composition of Tyrode solution.

Chemical compound	Concentration, g dm^{-3}
$NaCl_2$	8.00
$CaCl_2$	0.20
KCl	0.20
$NaHCO_3$	1.00
Na_2HPO_4	0.05
$MgCl_2$	0.01

Saturated calomel electrode (SCE) of KP-113 type served as the reference electrode. Platinum electrode of PtP–201 type was used as the auxiliary electrode. Anodic polarisation curves were registered and served as the ground for determination of typical factors describing resistance to electrochemical corrosion, i.e. corrosion potential E_{corr}, breakdown potential E_b, polarisation resistance R_p and corrosion current density i_{corr}.

Mechanical characteristics was determined by means of static uniaxial tension test on Instron testing machine of 1116 type. Next, the course of strain hardening curve of σ_p=f(ε) function was determined. In order to determine yield stress σ_p correctly, precise determination of actual stresses $\sigma_{r0.2}$, that wire samples were subject to during tensile test was a basic issue. For the load corresponding to the conventional yield stress, the value of actual stress $\sigma_{r0.2}$ was calculated from the formula (1):

$$\sigma_{r0.2} = \frac{R_{p0.2}}{\left[\sqrt{\frac{1}{1.002}} - \frac{\nu \cdot R_{p0.2}}{E}\right]^2}, \tag{1}$$

where:
$R_{p0.2}$ – conventional yield stress,
ν - Poisson ratio,
E –Young module.

Methodology of tracing out strain hardening curve j was described in studies [13,14].

3. Results

Table 4 presents results of electrochemical corrosion tests of wires made of X12CrNiMo17-12-2 steel, electrochemically polished, and Table 5 – polished and chemically passivated.

Table 4. Results of resistance to pitting corrosion of electrolytically polished wires (mean values).

Wire diameter d, mm	Logarythmic strain in drawing process ε	Corrosion potential E_{corr}, mV	Breakdown potential E_b, mV	Polarisation resistance R_p, kΩcm^2	Corrosion current density $i_{corr.}$, μA cm^{-2}	E_b-E_{corr}, mV
5.5	0	+73	+896	1520	0.017	+823
4.3	0.49	+7	+721	1030	0.025	+714
3.0	1.21	+21	+605	1010	0.022	+584
2.5	1.58	-14	+592	976	0.026	+606
2.2	1.83	-55	+526	930	0.027	+581
2.0	2.02	-51	+503	873	0.028	+554
1.85	2.18	-70	+419	972	0.030	+489
1.65	2.41	-70	+396	741	0.027	+469
1.45	2.67	-80	+362	711	0.036	+442
1.35	2.81	-123	+310	456	0.057	+433

Table 5. Results of pitting corrosion resistance tests of polished and chemically passivated wires (mean values).

Wire diameter d, mm	Logarythmic strain in drawing process ε	Corrosion potential E_{corr}, mV	Breakdown potential E_b, mV	Polarisation resistance R_p, kΩcm^2	Corrosion current density $i_{corr.}$, μA cm^{-2}	E_b-E_{corr}, mV
5.5	0	+187	+1200	4580	0.0015	+1387
4.3	0.49	+131	+1120	4492	0.0018	+1251
3.0	1.21	+47	+730	4220	0.002	+777
2.5	1.58	+41	+650	3970	0.0065	+691
2.2	1.83	-1	+600	3610	0.007	+601
2.0	2.02	-24	+582	3690	0.007	+606
1.85	2.18	-109	+521	3240	0.008	+630
1.65	2.41	-115	+480	3190	0.008	+595
1.45	2.67	-137	+440	3180	0.008	+577
1.35	2.81	-183	+392	2820	0.009	+575

Fig. 1 shows anodic polarisation curves of selected wire diameters, electrochemically polished, and Fig. 2 – wires that were polished and then passivated.

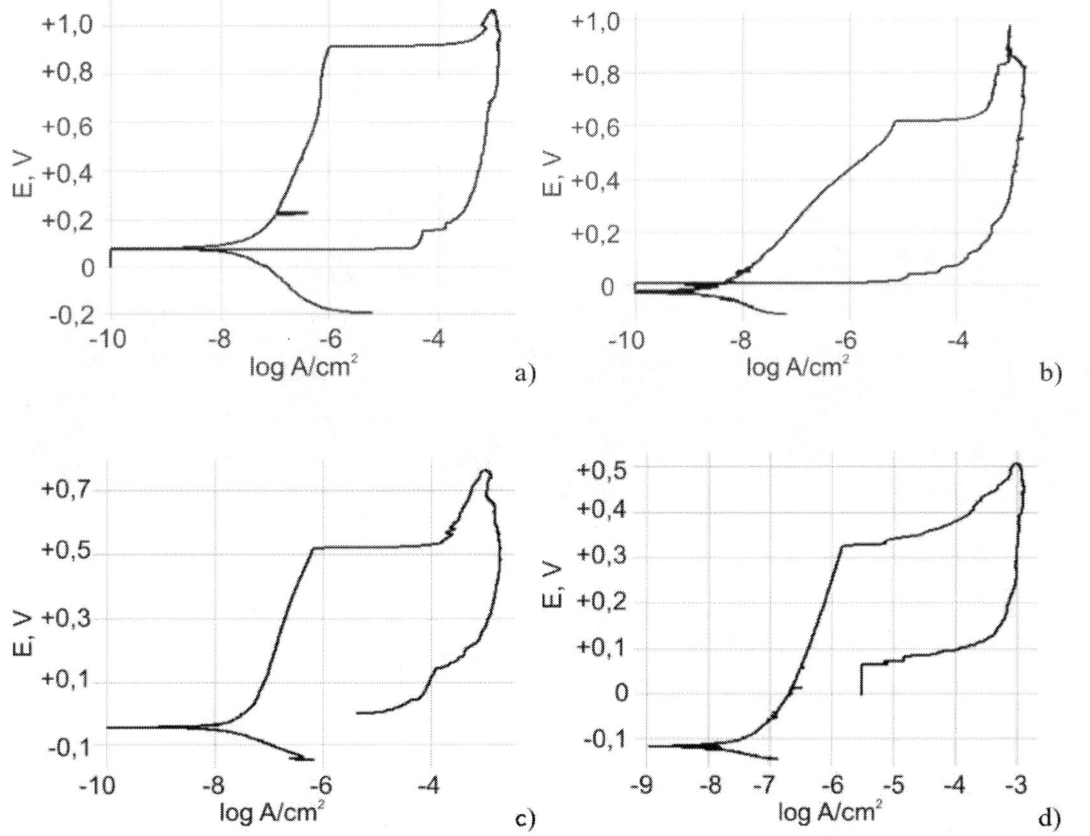

Figure 1. Anodic polarisation curves of electrochemically polished wire with diameter of: a) 5.5; b) 3.0; c) 2.0; d) 1.35 mm.

For wires that were electrochemically polished, mean value of corrosion potential of wire rod with diameter d_0=5.5 mm remains at the level of E_{corr}=+73 mV, whereas mean value of breakdown potential E_b = +896 mV. Corrosion current density and polarisation resistance are: i_{corr} = 0.017 μA cm^{-2} and R_p = 1520 kΩcm^2, respectively.

Average value of corrosion potential of polished wire with diameter of d = 1.35 mm is E_{corr} = -123 mV. Determined value of breakdown potential is E_b = +310 mV. Mean value of corrosion current density is i_{corr} = 0.057 μA cm^{-2}, and polarisation resistance R_p = 456 kΩcm^2.

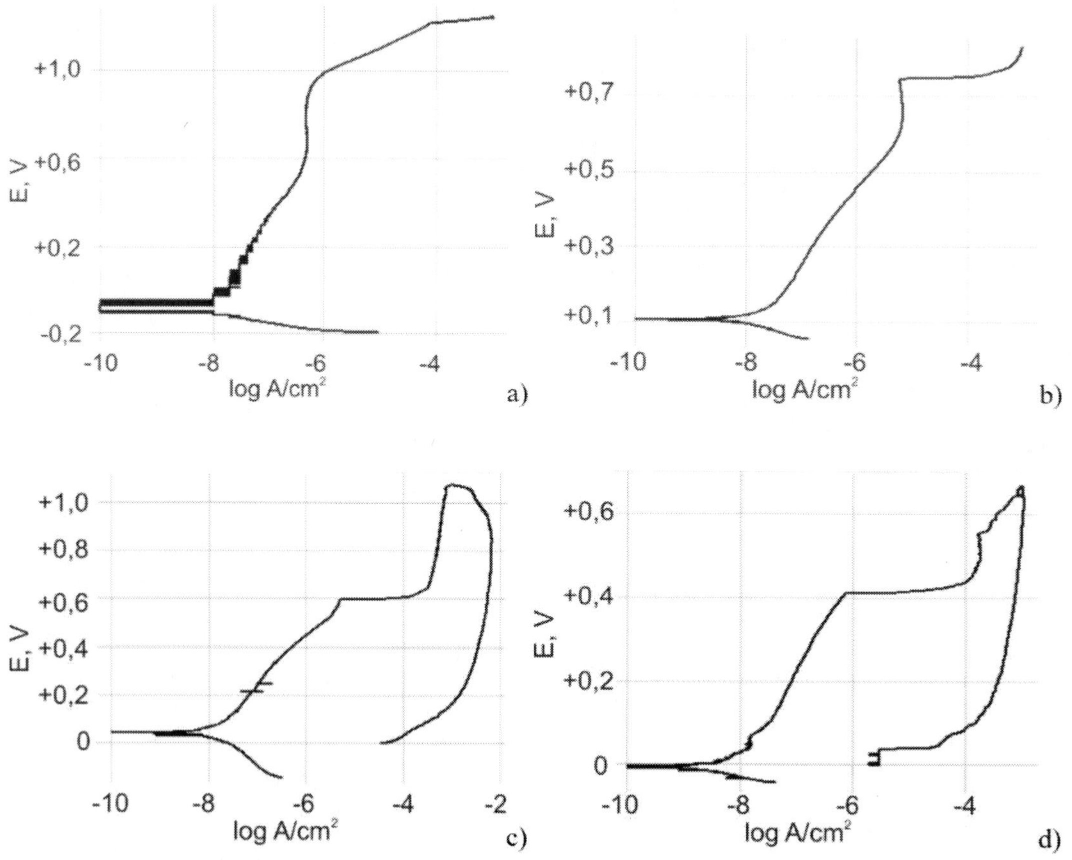

Figure 2. Anodic polarisation curves of electrochemically polished and passivated wire with diameter of: a) 5.5; b) 3.0; c) 2.0; d) 1.35 mm.

Mean value of corrosion potential of wire rod with diameter of d_0 = 5.5 mm after polishing and passivation remains at the level E_{corr} = +187 mV, whereas mean value of breakdown potential E_b = +1200 mV. Values of corrosion current density and polarisation resistance are: i_{corr} = 0.0015 μA cm^{-2} and R_p = 4580 kΩcm^{2,} respectively.
Mean value of corrosion potential of wire with diameter of d = 1.35 mm is E_{corr} = -183 mV. Determined value of breakdown potential is E_b = +392 mV. Mean corrosion current density is i_{corr} = 0.009 μA cm^{-2}, and polarisation resistance R_p = 2820 kΩcm^2.
Analysis of test results proves that supersaturated wire rod features the best corrosion resistance. Increase of strain applied in wire drawing process up to the value of ε=2.81 caused decrease of corrosion potential, breakdown potential and polarisation resistance. At the same time, there was increase in corrosion current density. These trends refer both to electrolytically polished wires and chemically passivated wires.
It was proved that with the increase of strain in drawing process, passive range decreases.
It was also proved that wire surface modification influenced substantially their resistance to electrochemical corrosion. Carried out tests prove explicitly substantial increase of corrosion parameters of wire as the results of polishing and chemical passivation when compared to single polishing.

Table 6 presents mechanical characteristics of wire after successive phases of drawing process.

Table 6. Strength properties of X2CrNiMo17-7-2 steel.

Diameter of wire, d_0 mm	ε	$R_{0,2}$ MPa	R_m MPa
5.5	0	224	533
4.3	0.49	854	972
3.0	1.23	1208	1353
2.5	1.58	1292	1433
2.2	1.83	1327	1521
2.0	2.02	1327	1550
1.85	2.18	1406	1579
1.65	2.41	1408	1645
1.45	2.67	1462	1674
1.35	2.81	1429	1726

Actual stress, determined on the ground of tensile test, corresponding to yield stress, were used to draw strain hardening curve of the tested wires and to determine mathematical form of yield stress function. The curve was approximated by means of function of $\sigma_p=\sigma_{p0}+C\varepsilon^n$ type, that takes the value of strain for initial state (i.e. for wire after heat treatment, designed for drawing) into account. Fig. 3 presents strain hardening curve of the tested wires. Mathematical form of yield stress function for the tested steel is as follows (2):

$$\sigma_p=224.8 + 878\varepsilon^{0.4}. \tag{2}$$

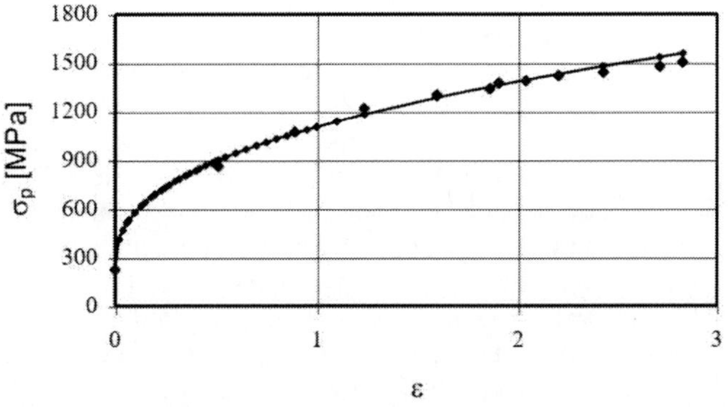

Figure 3. Strain hardening curve for wire made of X2CrNiMo17-12-2 steel.

The results of potentiodynamic tests were then subject to statistical analysis. It was determined that there is a significant relation between corrosion properties and strain in drawing process. The tests that were carried out enabled to select functions depicting the change of polarisation resistance, depending on strain applied in drawing process. They are as follows (3) and (4):

$$R_p = -267.05\varepsilon_c+1381.2 \text{ (electrochemically polished wires),} \tag{3}$$
$$R_p = -631.65\varepsilon_c+4785.6 \text{ (electrochemically polished and passivated wires).} \tag{4}$$

In both cases the level of significance was $p<0.05$.

Fig. 4 presents curves obtained on the ground of the selected corrosion tests results, namely the change of polarisation resistance as the function of strain applied in drawing process.

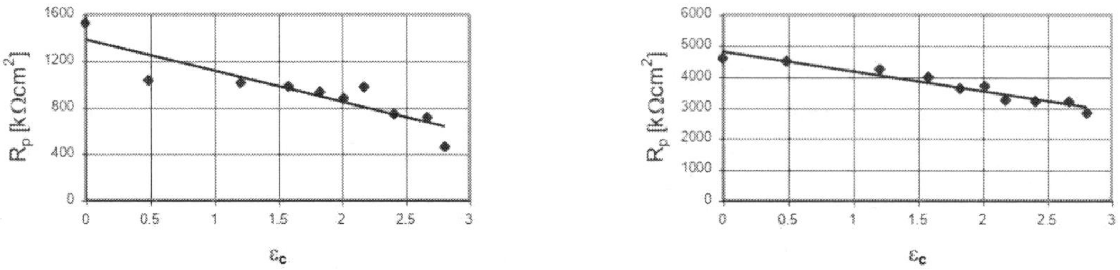

a) b)

Figure 4. The relation between polarisation resistance and strain in the process of wire drawing: a) electrochemically polished, b) polished and passivated.

Presented curves in regular way reflect the results obtained in the experiment. The biggest differences of polarisation resistance can be observed for wire with initial diameter.
Presented curves enable to deduce polarisation resistance of wires subject to surface treatment. Polarisation resistance value proves that chemical passivation of wires for orthopaedics, made of X2CrNiMo17-12-2 steel, is justified.

4. Summary

Potentiodynamic tests carried out in Tyrode solution provided information about resistance to electrochemical corrosion of wires made of stainless steel X2CrNiMo17-12-2 for bone implants production. Comparative analysis of anodic polarisation curves proved that strain applied in drawing process has significant influence on wire corrosion parameters. With increase of strain in drawing process, decrease of corrosion potential was observed as well as decrease of breakdown potential and polarisation resistance. It was proved that corrosion current density and corrosion rate increased. Wire rod with diameter $d_0 = 5.5$ mm featured the best corrosion characteristics.
Corrosion resistance of wires also depends on their surface preparation. The tests confirmed that wire surface modification, which results in improved physical and chemical characteristics of the surface layer, is justified. Potentiodynamic measurements that were carried out showed favourable influence of electrolytic polishing, and then chemical passivation, applied after grinding process, on wire corrosion. The best corrosion characteristics in Tyrode solution can be attributed to wires with electrochemically polished and then chemically passivated surface.
Executed work has a direct application effect because it links technological process of wire production for orthopaedics with their corrosion properties. As wires for orthopaedics implants should feature suitable mechanical characteristics for the respective application, and they most of all condition drawing technology, it is not possible to influence their corrosion resistance by means of strain applied in cold working. It was proved that in each case, irrespective of work hardening, presence of pits can be observed. It proves low resistance of steel X2CrNiMo17-12-2 to electrochemical corrosion in the environment of body fluids. Test results prove that application of protective layers on wires made of chromium – nickel – molybdenum steel resistant to corrosion is essential.
Carried out tests of mechanical characteristics enabled to determine the course of strain hardening curves and select yield stress function of drawn wires made of X2CrNiMo17-12-2 steel. Next, statistical analysis of results of electrochemical corrosion resistance tests enabled to obtain functional

relations showing the influence of strain in drawing process on change of corrosion parameters. The study only presents examples of such relations form polarisation resistance, but they can also be determined for other factors.
It must be mentioned that given functional relations are not universal and only refer to wires made of X12CrNiMMo17-12-2 steel tested in Tyrode solution. When other material is used for wire production, or when products are designed for different purpose, it is necessary to carry out similar tests of wire made of that material in other environment simulating human body fluids.
Presented aspects are extremely important for production engineers dealing with drawing process designing, because the issue of correct description of material workability is tightly connected with selection of optimum parameters of technological process of wire production. This issue is also important for manufacturers of bone implants of wire and wire products, as having the respective curves or functions, it is possible to predict in advance corrosion characteristics of wire with required strength, drawn with the respective strain.

Acknowledgments
This project was financed from the funds of National Science Centre in Cracow.

References
[1] Marciniak J 2002 *Biomateriały* (Gliwice Wydawnictwo Politechniki Śląskiej).
[2] Krauze A, Ziębowicz A and Marciniak J 2005 *J. Mater. Proces. Techn.* **162-163** 209.
[3] Grosman F and Przondziono J 2004 *Obróbka Plastyczna Metali* **15** 5.
[4] Przondziono J, Halaczek D and Szymszal J 2007 *Arch. Civil Mech. Eng.* **7** 85.
[5] Przondziono J and Walke W 2009 *Arch. Mater. Sci. Eng.* **35** 21.
[6] Walke W and Przondziono J 2010 *Solid State Phenomena* **165** 404.
[7] Przondziono J 2011 *Hutnik - Wiadomości Hutnicze* **78** 663.
[8] Łuksza J, Skołyszewski A, Witek F and Zachariasz W 2006 *Druty ze stali i stopów specjalnych* (Warszawa: WNT).
[9] Kaczmarek M, Walke W and Kajzer W 2007 *Archiv. Mater. Sci. Eng.* **5** 273.
[10] Walke W and Przondziono J 2010 *J. Achiev. Mater. Manufact. Eng.* **39** 27.
[11] Walke W and Przondziono J 2011 *Metalurgija (Metallurgy)* **50** 201.
[12] Ziębowicz A, Walke W, Barucha-Kepka A nad Kiel M 2008 *J. Achiev. Mater. Manufact. Eng.* **27** 154.
[13] Przondziono J and Walke W 2011 *IOP Conf. Series: Mater. Sci. Eng.* **22** 012018 1.
[14] Walke W and Przondziono J 2011 *Eng. Biomater.* **14** 145.
[15] Walke W and Przondziono J 2012 *Metalurgija (Metallurgy)* **51** 234.
[16] ASTM G61 – 86 2009 Standard Test Method for Conducting Cyclic Potentiodynamic Polarization Measurements for Localized Corrosion Susceptibility of Iron-, Nickel-, or Cobalt-Based Alloys.
[17] ASTM F2129 - 08 Standard Test Method for Conducting Cyclic Potentiodynamic Polarization Measurements to Determine the Corrosion Susceptibility of Small Implant Devices.

IOP Conf. Series: Materials Science and Engineering **35** (2012) 012016 doi:10.1088/1757-899X/35/1/012016

Corrosion resistance of premodeled wires made of stainless steel used for heart electrotherapy leaders

J Przondziono[1], W Walke[2], R Młynarski[3] and W Szatka[1]
[1]Department of Technology Materials, Silesian University of Technology, Krasińskiego 8, 40-019 Katowice, Poland
[2]Department of Biomaterials and Medical Devices Engineering, Silesian University of Technology, Akademicka 16, 44-100 Gliwice, Poland
[3]Department of Electrocardiology, Medical University of Silesia, Ziołowa 45/47, 40-635 Katowice, Poland

E-mail: joanna.przondziono@polsl.pl

Abstract. The purpose of the study is to evaluate resistance to electrochemical corrosion of wire made of X10CrNi18-8 stainless steel designed for use in cardiology treatment. The influence of strain formed in the premodeling process and methods of wire surface preparation to corrosive resistance in artificial plasma solution were analysed. Wire corrosion tests were carried out in the solution of artificial plasma. Resistance to electrochemical corrosion was evaluated on the ground of recorded curves of anodic polarization by means of potentiodynamic method. Potentiodynamic tests carried out enabled to determine how the resistance to pitting corrosion of wire changes, depending on strain formed in the premodeling process as well as on the method of wire surface preparation. For evaluation of phenomena occurring on the surface of tested steel, electrochemical impedance spectroscopy (EIS) was applied. Deterioration of corrosive properties of wire along with the increase in the formed strain hardening was observed.

1. Introduction

Tests within the range of biomedical engineering carried out currently focus mainly on adequate selection of physical and chemical properties of materials that shall meet extremely restrictive criteria of their biological evaluation. It also refers to materials that will be used for production of surgical instruments which are used among other things for the correct location of implants in the human body or electrodes during heart electrotherapy. One of such instruments are guide wires used for implantation of heart pacemakers. The first external heart pacemaker was implanted in 1958 by a Swedish heart surgeon Ake Senning in cooperation with its manufacturer Rune Elmeqvist. Electrodes of the device were then placed on the epicard wall, not inside the heart. It was a ground-breaking moment in the development of electrotherapy whose treatment efficiency was proved both by scientific research as well as clinical practice [1-3]. The level of the development in the area of microelectronics is nowadays so high that it enables to make use of devices that are substantially smaller, with substantially greater programming options, and electrodes of pacemakers have become much thinner and more durable. In order to provide proper electrical stability and correction of conduction disturbances in a way that enables to obtain electrical operation of the heart as similar to the regular condition as possible, the tests are still being carried out to improve and modify the existing equipment and instruments.

Guide wires, the purpose of which is to insert endocavitary electrodes, are of crucial importance [3]. Wires made of stainless steel with diverse diameters and mechanical properties are mostly used for this purpose. Currently on the market there are many such guide wires that differ not only as far as material used is concerned, but most of all because they feature different forms and geometrical features. Such a huge variety of geometries is caused more by patent and market issues than by real clinical considerations. They haven't been normalised so far. Each company manufactures guide wires using their own catalogue series of types, determining their typical features and properties. However, clinical practice shows such limitations connected with currently used materials for guide wires. They are first and foremost connected with biocompatibility. Material inserted into blood-vascular system cannot cause changes to electrolyte composition, irreversible damage to protein structure, release of morphotic blood elements and it also should not initiate the process of coagulation, toxic and immunological reactions. Thus, the interest of manufactures in adequate surface treatment is increasing, in order to limit the occurrence of those unfavourable phenomena [4].

Guide wires used commercially in electrotherapy are usually in the form of wires and are made of stainless steel of 18-8 type [5]. The analysis of the course of atrial lead implantation shows that guide wire prior to its implantation is deformed permanently. Therefore, in this study the tests were carried out that aimed at determination of influence of plastic strain of the wire on its corrosion resistance in blood environment. In order to simulate working conditions guide wires were deformed in the angle bend test by the angles α: 30°, 60°, 90°, 120° and 150°, respectively, on the radius of $r = 3$ mm. The tests were carried out by means of potentiodynamic method which enabled to determine typical parameters describing resistance corrosion of the guide wire. Parameters obtained in the measurements were compared with non-deformed guide wire, which made the ground for the assessment of applicability of stainless steel X10CrNi18-8 for such type of medical product.

2. Methods and procedures

Guide wires, commercially used in hospitals, made of stainless steel of X10CrNi18-8 type, in the form of wire with diameter of $d = 0.4$ mm and length $l = 680$ mm – Fig. 1, were used for the tests. Prior to the test, guide wires were sterilised with pressurised water steam in autoclave at the temperature $T = 134$°C for $t = 12$ min.

Potentiodynamic tests began with determination of open circuit potential E_{OCP} when there was no current. Then, anodic polarisation curves were registered, and the measurement commenced with the potential equal $E_{start} = E_{OCP} - 100$ mV. Potential changed in anodic direction at the rate of 1 mV/s. When the density of anodic current reached $i = 1$ mA cm^{-2} , polarisation direction was changed. This way return curve was registered. Registered curves made the ground for determination of typical parameters describing resistance to pitting corrosion, i.e.: corrosion potential E_{corr} (mV), breakdown potential E_b (mV), polarisation resistance R_p (k$\Omega \cdot$cm^2), corrosion current density i_{corr} (μA cm^{-2}). Measurements were carried out with the use of measurement system VoltaLab®PGP201. Saturated calomel electrode NEK served as the reference electrode, whereas platinum electrode served as the auxiliary electrode.

In order to obtain information about physical and chemical characteristics of samples made of steel X10CrNi18-8, tests that employed electrochemical impedance spectroscopy were carried out. Measurements were made with application of testing system Auto Lab PGSTAT 302N equipped with FRA2 module (Frequency Response Analyser). The tests began with determination of open circuit potential E_{OCP} when there was no current. The applied measurement system enabled to carry out tests in frequency range $10^4 \div 10^{-3}$Hz. Amplitude of square-wave voltage of activation signal was

10 mV. The tests enabled to determine impedance spectra of the system and match obtained measurement data to the equivalent system. Impedance spectra of tested system have been presented in the form of Nyquist diagrams for various frequencies and as Bode diagrams. Obtained EIS spectra were interpreted after matching to the equivalent electrical system by means of the smallest square method. The tests were carried out in the solution of artificial blood plasma at the temperature of T = 37±1°C – Table 1.

Table 1. Chemical composition of the artificial plasma according to PN – EN ISO 10993-15 standard.

Components	NaCl	$CaCl_2$	KCl	$MgSO_4$	$NaHCO_3$	Na_2HPO_4	NaH_2PO_4
G dm^{-3} distilled water	6.8	0.2	0.4	0.1	2.2	0.126	0.026

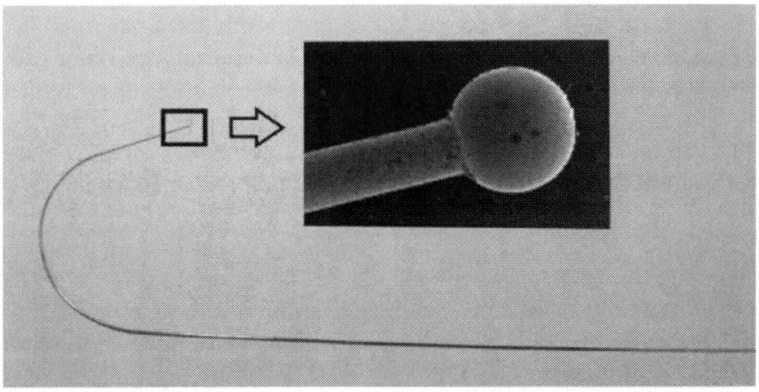

Figure 1. Guide wire with the tip profiled at the angle α = 150°.

3. Results

On the ground of the carried out potentiodynamic tests it was determined that corrosion potential for non-deformed guide wires established at E_{corr} = -73 mV – Fig. 2a. Polarisation brought about rapid increase of anodic current intensity at the potential E_b = +760 mV – Fig. 2a and 3. When the value of anodic current anodic current reached 1 mA/cm^2, the direction of samples polarisation was changed, which enabled to register the return curve. Anodic polarisation curve, registered in such a way, featured the presence of hysteresis loop which proved the presence of pitting corrosion. Moreover, on the ground of Stern method, polarisation resistance and corrosion current density were determined, and they were R_p = ÷ 69.5 kΩ cm^2 and i_{corr} = 0.374 μA cm^{-2} respectively – Table 2.

Table 2. Results of corrosion resistance.

Bend angle α,°	E_{OCP} (mV)	E_{corr} (mV)	R_p (kΩ cm^2)	$i_{corr.}$ (μA cm^{-2})	E_b (mV)	E_b - E_{corr} (mV)
0	-95	-73	69.5	0.374	+760	833
30	-102	-108	84.2	0.309	+725	833
60	-117	-24	95.5	0.272	+562	586
90	-129	-81	67.9	0.383	+517	598
120	-144	-68	76.2	0.341	+403	471
150	-148	-65	77.2	0.337	+380	445

Next, samples were analysed after application of strain at the angle of α: 30°, 60°, 90°, 120° and 150°, respectively. Corrosion potential fell in the range of E_{corr} = -108 ÷ -24 mV – Fig. 2b-2f, whereas values of breakdown potential were smaller than the values obtained for non-deformed guide wires and fell within the range of E_b = +380 ÷ + 725 mV – Fig. 2b-2f and 3. Next, values of polarisation resistance and corrosion current density were differentiated depending on bend angle. These values are, respectively, in the range for R_p = 67.9 ÷ 95.5 kΩ cm^2, i_{corr} = 0.272 ÷ 0.383 μA cm^{-2} – Table 2.

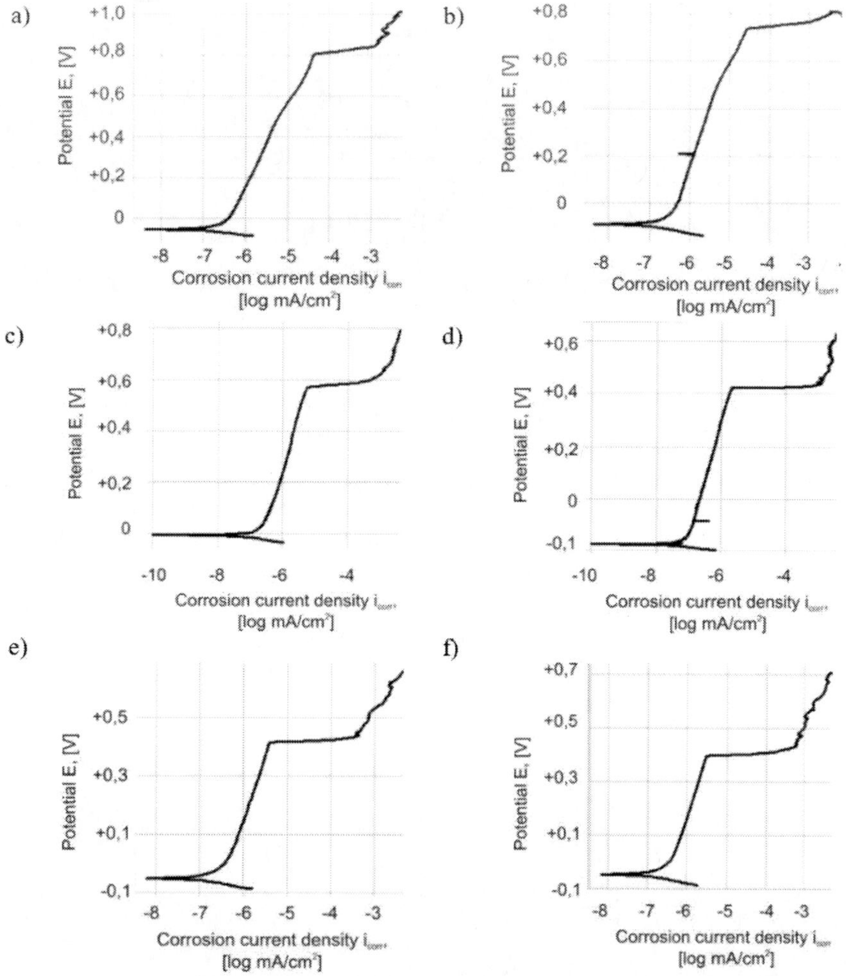

Figure 2. Anodic polarisation curves determined for guide wire with bend angle: a) α = 0°, b) α = 30°, c) α = 60°, d) α = 90°, e) α = 120°, f) α = 150°.

IOP Conf. Series: Materials Science and Engineering **35** (2012) 012016 doi:10.1088/1757-899X/35/1/012016

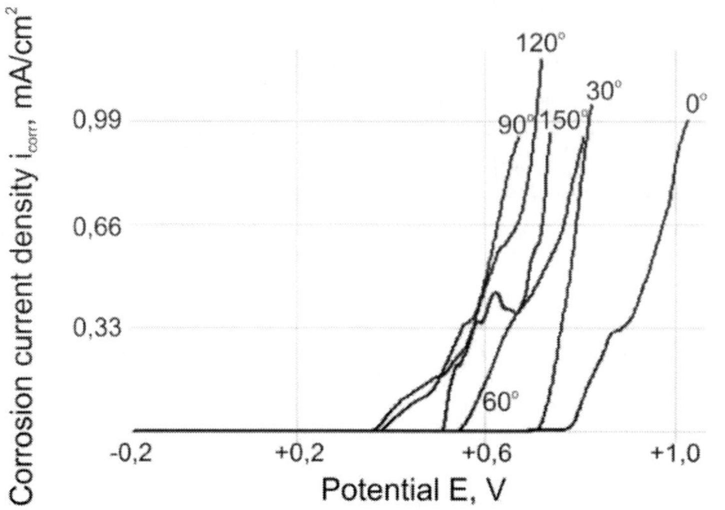

Figure 3. Value of perforation potential for the angle of guide wire bending angle: a) α = 0°, b) α = 30°, c) α = 60°, d) α = 90°, e) α = 120°, f) α = 150°.

Exemplary impedance spectra registered for samples made of X10CrNi18-8 steel, non-deformed and deformed in the bend test by the angle of 150°, are presented in Fig. 4 and 5. Nyquist diagrams determined for samples prepared in such a way show fragments of large incomplete circles that reflect typical impedance response for thin oxide layers – Fig. 4a and 5a. Maximum values of phase angle, in the wide range of frequencies presented in Bode diagrams irrespective of bend angle, are similar and equal $\theta \approx 75°$. Slopes $\log|Z|$ within the whole range of frequency changes are close to -1, which proves capacitive character of the passive layer – Fig. 4b and 5b. Next, high values of impedance $|Z| > 10^6$ Ωcm^2 in the range of the smallest frequencies prove good dielectric and protective characteristics of passive layers created on the surface of the samples made of X10CrNi18-8 steel, non-deformed and deformed – Table 3.

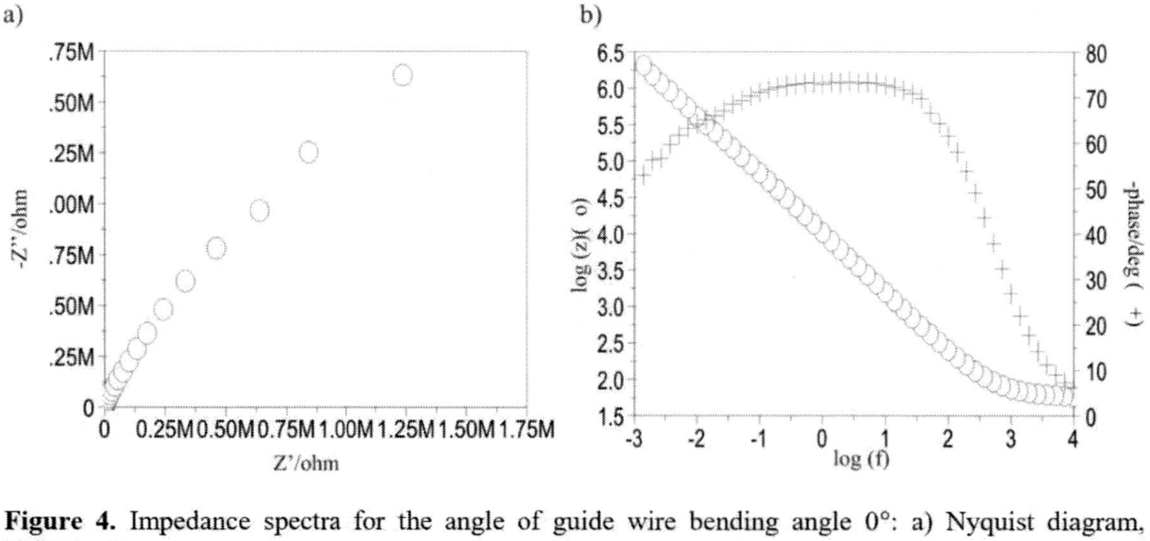

Figure 4. Impedance spectra for the angle of guide wire bending angle 0°: a) Nyquist diagram, b) Bode diagram.

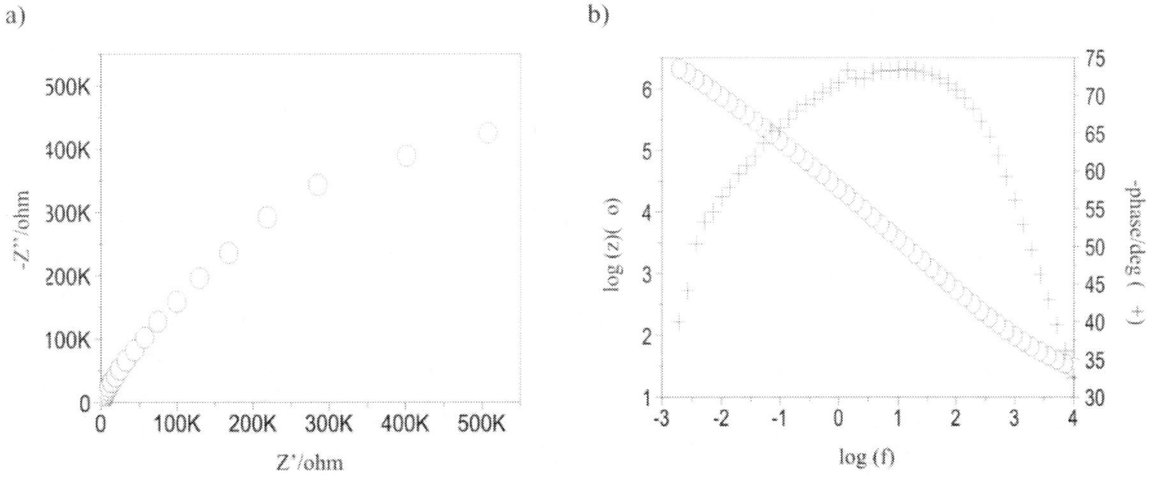

Figure 5. Impedance spectra for the angle of guide wire bending angle 150°: a) Nyquist diagram, b) Bode diagram.

Characteristics of impedance of phase boundary: electrode – passive layer – solution was made through approximation of experiment data by means of electric equivalent circuit model – Fig. 6. Values of typical parameters describing electrical characteristics of non-deformed samples slightly differed from values determined for deformed samples – Table 3. It was determined that with the increase of bend angle, R_p decreases. Measurement results of deformed and non-deformed were matched to the simplest model of oxide layer, i.e. such that consists of a parallel CPE (Constant Phase Element) connected with oxide layer resistance (passive) R_p and resistance at high frequencies, which can be attributed to electrolyte resistance R_s [6-9]. In the electrical equivalent system resistor R_p and CPE represent, respectively, resistance and passive oxide layer capacity, created on the alloy surface. Whereas resistor R_s represent resistance of artificial blood plasma in which the test were made – Table 3.

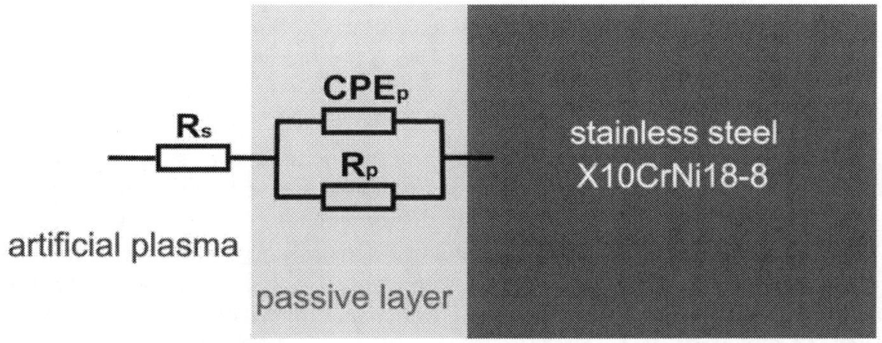

Figure 6. Electrical model of equivalent circuit for X10CrNi 18-8 – passive layer – artificial plasma.

Mathematical impedance model of the X10CrNi 18-8 – passive layer – artificial plasma system is presented in equation (1):

$$Z = R_s + \frac{1}{\frac{1}{R_p} + Y_0 (j\omega)^n}. \tag{1}$$

Table 3. Results of electrochemical impedance spectroscopy.

Bend angle α, °	E_{OCP} (mV)	R_s (Ωcm²)	R_p (MΩcm²)	CPE	
				Y_0, $\Omega^{-1}cm^{-2}s^{-n}$	n
0	-90	14	4.203	0.6453e-4	0.77
30	-112	15	3.668	0.5476e-4	0.77
60	-122	12	2.798	0.6345e-4	0.78
90	-118	13	1.890	0.4456e-4	0.79
120	-138	15	1.450	0.3535e-4	0.79
150	-144	15	1.023	0.3245e-4	0.80

Conclusions

Popularity of heart electrotherapy leads directly to the increase of patients who undergo such treatment. Success of electrotherapy depends on a range of factors, and one of them is proper selection of instruments and auxiliary materials. Of crucial importance for proper realisation of cardiological pacemaker implantation are guide wires whose purpose is to insert endocavitary electrodes. Tests regarding guide wires manufacturing technology are carried out in order to obtain products with proper performance characteristics, including also their improved biocompatibility.

Clinical tests analysis shows that during electrode implantation guide wire is permanently deformed prior to treatment performed by the operator. That is why the tests of its susceptibility to plastic deformation in working conditions were made. Samples were deformed in bend test by the angle α: 30°, 60°, 90°, 120° and 150°. Analysis of X-ray shadow proves that guide wire geometry, and to be precise – its tip, are subject to deformation that generally does not exceed the angle α = 120°. However, in order to make a full evaluation, they were deformed within the range exceeding that value. For samples made of X10CrNi 18-8 steel it was proved that the increase in deformation angle value decreases breakdown potential value E_b by ca. 380 mV, which is an unfavourable phenomenon. Determined passive range E_b - E_{corr} also decreased with the increase of strain from 833 mV for non-deformed condition to the value of 445 for the condition after deformation by the angle of 150°. It proves that corrosion resistance of guide wire decreased with the increase of deformation angle – Fig. 3 and Table 2. Registered values of open circuit potential, corrosion potential or polarisation resistance did not differ substantially for deformed and non-deformed guide wires.

In order to make a more detailed evaluation of all phenomena taking place on the surface of deformed samples, the test in which electrochemical impedance spectroscopy was employed were made next. Selection of this method enabled to characterise impedance of phase boundary: biomaterial – passive layer – solution, by means of approximation of impedance data using electrical equivalent system model. Carried out EIS analysis enabled to determine impedance spectra of the tested system and to match data to equivalent system that consisted of parallel system of CPE connected with layer resistance R_p and residuary resistance R_s which at high frequencies is attributed to ohm of the artificial plasma solution. Determined values of admittance (Z^{-1}) and n coefficient for samples after exposure in the plasma solution showed that intrinsically created passive layer was not destroyed in contact with artificial blood plasma. The only phenomenon that was observed was decrease of resistance R_p with increase of bend angle – Table 3.

Therefore, it is believed that production engineers should focus their work on modification of guide wires surface in order to increase their resistance to electrochemical corrosion. Creation of suitable protective layers would enable to eliminate unfavourable phenomena in the blood and vascular system and substantial improvement of biocompatibility of used steel wires.

Acknowledgments
This project was financed from the funds of National Science Centre in Cracow.

References

[1] Luderitz B 2002 *J. Cardiovasc. Electrophys.* **13** (Suppl. 1) 2.
[2] Elmqvist R and Senning A 1959 *Proc. of the Second International Conference on Medical Electronics (Paris),* (London: Iliffe & Sons) 253.
[3] Kargul W, Młynarski R and Piłat E 2005 *Chirurgia Polska* **7** 267.
[4] Paszenda Z 2005 *Kształtowanie własności fizykochemicznych stentów wieńcowych ze stali Cr-Ni-Mo do zastosowań w kardiologii zabiegowej* (Gliwice: Wydawnictwo Politechniki Śląskiej), in polish.
[5] Przondziono J, Młynarski R, Szala J and Kur A 2010 *Eng. Biomater.* **13** 56.
[6] Walke W and Przondziono J 2011 *Metalurgija (Metallurgy)* **50** 201.
[7] Walke W and Przondziono J 2010 *Diffusion and Defect Data Pt.B: Solid State Phenomena* **165** 404.
[8] Basiaga M, Paszenda Z and Walke W 2011 *Przegląd Elektrotechniczny* **12b** 12.
[9] Łosiewicz B and Budniok A 2003 *Corrosion of Protection* **11** 49.

IOP Conf. Series: Materials Science and Engineering **35** (2012) 012017 doi:10.1088/1757-899X/35/1/012017

Study on Cracking Process of Power Boiler Element

K Mutwil, M Cieśla,
Department of Technology Materials, Silesian University of Technology, Krasińskiego 8, 40-019 Katowice, Poland

E-mail: kazimierz.mutwil@polsl. pl marek.ciesla@polsl.pl

Abstract:This paper presents the study of the reasons for cracking of spray steam attemperator. The element being analysed is a temperature controller of steam for turbine; its proper functioning has considerable impact on operation of power generating set of boiler and turbine. Damage to the steam attemperator may affect the durability of neighbouring components, replacement or repair of which results in a long-term outage of power generation unit.
The reason for attemperator cracking has been determined on the basis of strength calculations and micro-and macro-analysis of fracture surfaces. The conducted tests were the basis for modification of attemperator design, which will ensure reliable functioning of the analysed element.

1. Introduction

Pressure equipment in power and heat generation plants is designed for a certain period of operation amounting usually to 100 000 or 200 000 operating hours. Computing life-time assumed for engineering works is deemed a safe period of equipment operation. Nevertheless, in practice there is a large number of plants, which, despite a considerable excess of computing life-time, are still operating safely, and on the other hand, there are numerous examples of damages occurred to the plant components, which resulted in their withdrawal from use prior to expiration of the computing life-time [1, 2].

Complexity of interactions occurring between various design, process, manufacturing and operation-related factors, which have not been considered in engineering calculations, causes that the operation time of many components of power generation plants cannot be exactly determined or its accurate determination is very difficult [3].

For unbiased life-time evaluation, extensive theoretical background and engineering practice, which enable reasonable selection of research tools and methods and proper interpretation of their results, are necessary. Current research trends in power generation sector should aim at wider and wider implementation of PC-aided techniques, in which the results of equipment operation monitoring and historical data of plant operation as well as material properties after long-term operation shall be properly used [4, 5].

In this paper the reasons for cracking of spray tube in S2 steam attemperator installed in a power boiler are discussed. The results of fracture investigations and microstructure analysis of damaged element were used in the researches. A numerical modelling was used for determination of the load of steam attemperator components in the operating conditions.

2. Object of the Study

In a power boiler, transition of water to superheated steam runs in several stages in the so called "steam superheaters". Steam attemperators are provided between the individual stages of steam superheaters. Basic function of an attemperator is providing the required temperature of steam supplied to the turbine irrespective of actual boiler load. Temperature control of the steam is mostly disregarded in small boiler units, it is used as required in industrial boilers, and it is always used in power boilers. In the 60-ties of 20th century, keeping a constant steam temperature was required for the range of boiler load of 75% to 100% while nowadays this requirement generally applies to the boiler load ranging from of 50 ÷ 100%.

Steam temperature in power boiler is controlled by means of various attempering systems. Spray attemperators are most commonly used for steam temperature control. Overall geometry features and main dimensions of attemperator are given on Fig. 1.

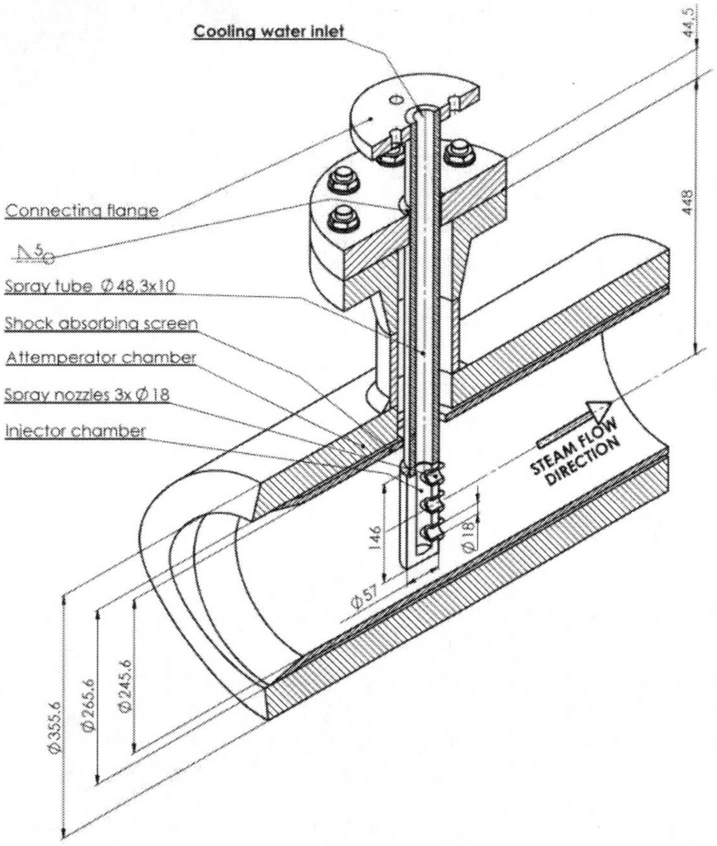

Figure 1. Construction and main dimensions of spray steam attemperator

A steam of a rated temperature of 520^0C and pressure of 20.6 MPa flows through an attemperator chamber with a mass-flow rate of 79.8kg/s. Pressure of injected water is of approx. 21 MPa, temperature of outside wall of spray tube is of approx. 350^0C, and injected water temperature amounts to approx. 250-270°C. The injector chamber and the spray tube are made of steel grade 11CrMo9-10 and 13CrMo4-5, respectively.

In the steam attemperators of the construction shown on Fig.1, numerous repeated cracks of spray tube in the area of its connection with a connecting flange have been found during operation. Location of cracks and morphology of spray tube fracture are shown on Fig. 2.

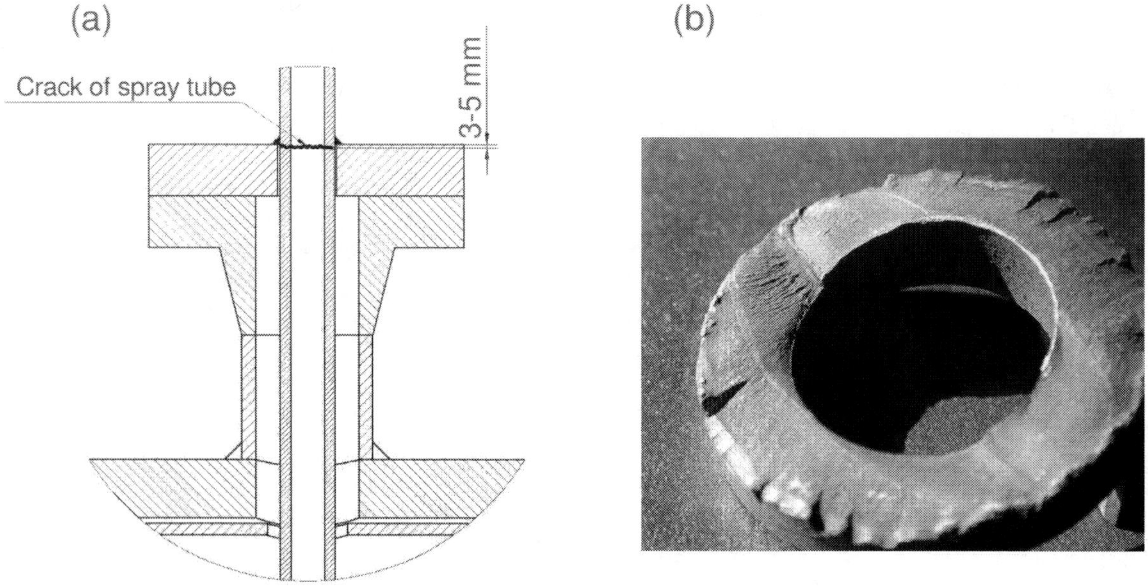

Figure 2. Location of cracks (a) and morphology of spray tube fracture (b)

3. Analysis of Spray Tube Cracking Process

One of the metallographic test methods, which enables obtaining essential information of the causes, nature and propagation of cracks is the analysis of crack (fracture) surface - the so called "fracture investigation". In the fracture investigations, the following two types of tests, which, when in combination, provide comprehensive information of the nature and the reasons of structure cracking, are used:

- **structural tests**, which determine the relations between the fracture particulars and the microstructure and defects of the material.
- **mechanical tests**, which determine correlation between the fracture particulars and the operating conditions of the object; the manner of load application, object temperature, geometry as well as other factors are taken into account in these tests.

3.1. Structural Tests

Macroscopic analysis of injector fracture revealed the cracking process had been initiated at injector's outside surface and crack propagation run along the plane perpendicular to spray tube axis (Fig. 2 and 3). Two fatigue centres located in "A" and "B" areas at the opposite sides of the spray tube were found on the fracture. A symmetric initiation of cracking process proves the process was running in conditions of oscillatory bending [6, 7].
Macroscopic observations of crack surface revealed also the occurrence of typical fatigue line patterns (Fig. 4, 5, 6a). When observing the fracture in a scanning microscope, fatigue bands and replicated cracks, which are typical features of fatigue scrap morphology, were found on the fracture surface (Fig. 6b).

IOP Conf. Series: Materials Science and Engineering **35** (2012) 012017 doi:10.1088/1757-899X/35/1/012017

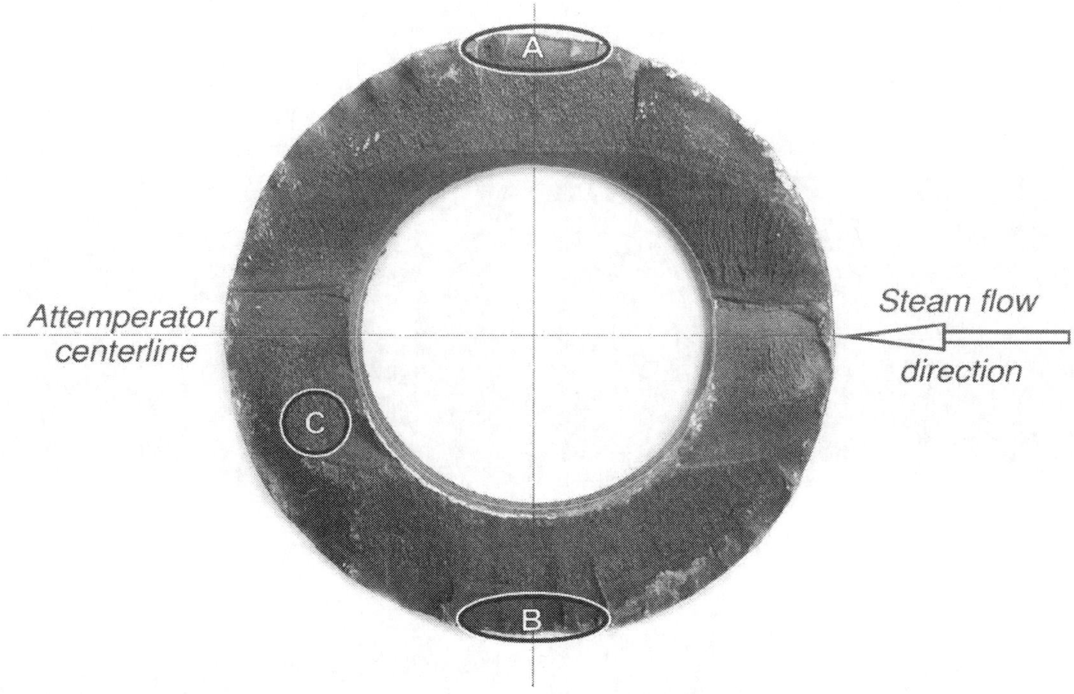

Figure. 3. Macrostructure of fatigue scrap

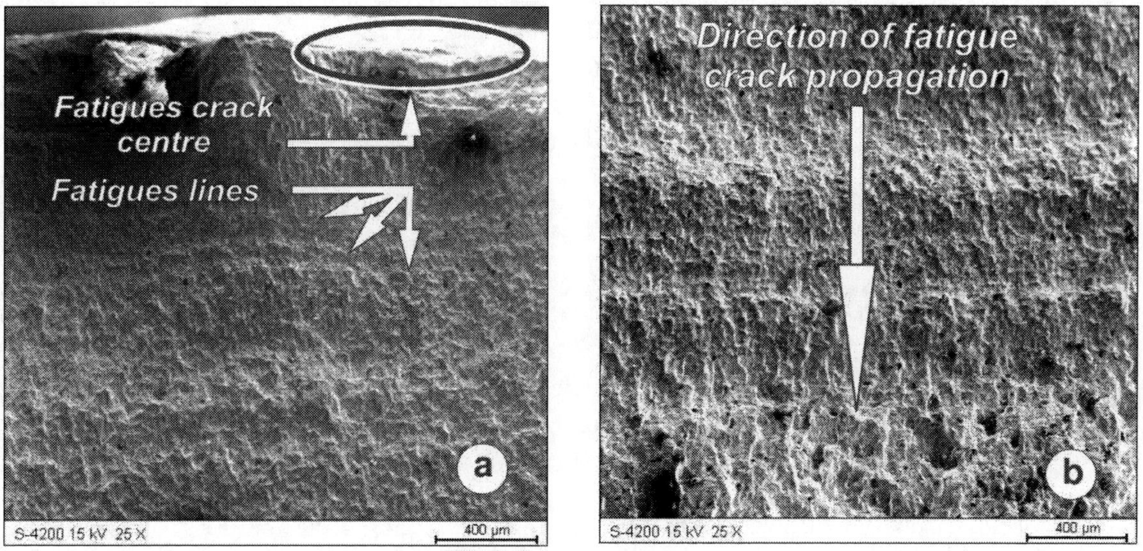

Figure 4. Fatigue scrap of spray tube ("A" area on Fig. 3); (a) scrap morphology in area around the crack centre, (b) fatigue line pattern

IOP Conf. Series: Materials Science and Engineering **35** (2012) 012017 doi:10.1088/1757-899X/35/1/012017

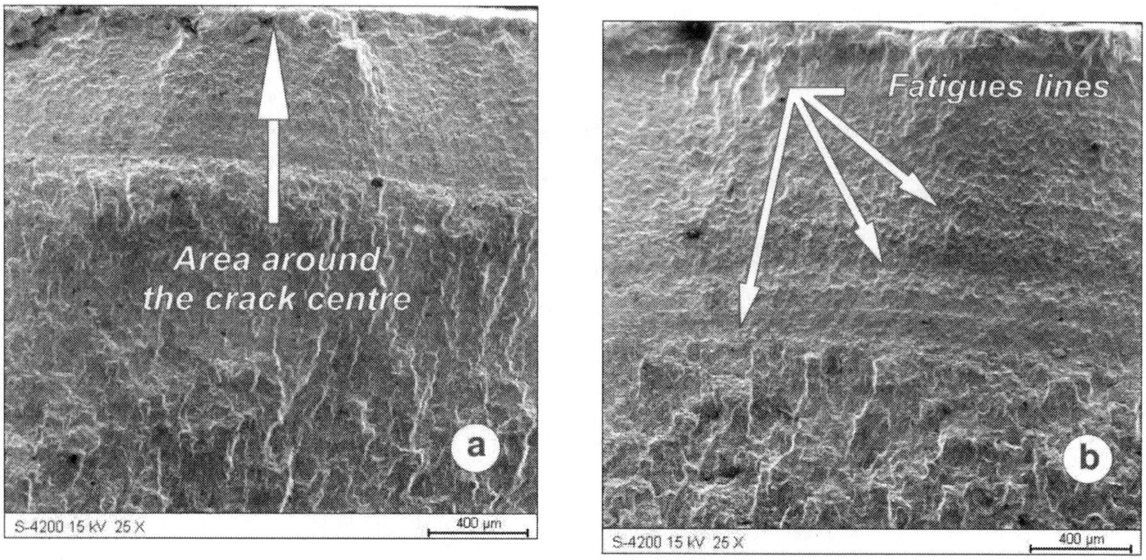

Figure 5. Fatigue scrap of spray tube ("B" area on Fig. 3); (a) scrap morphology in area around the crack centre, (b) fatigue line pattern

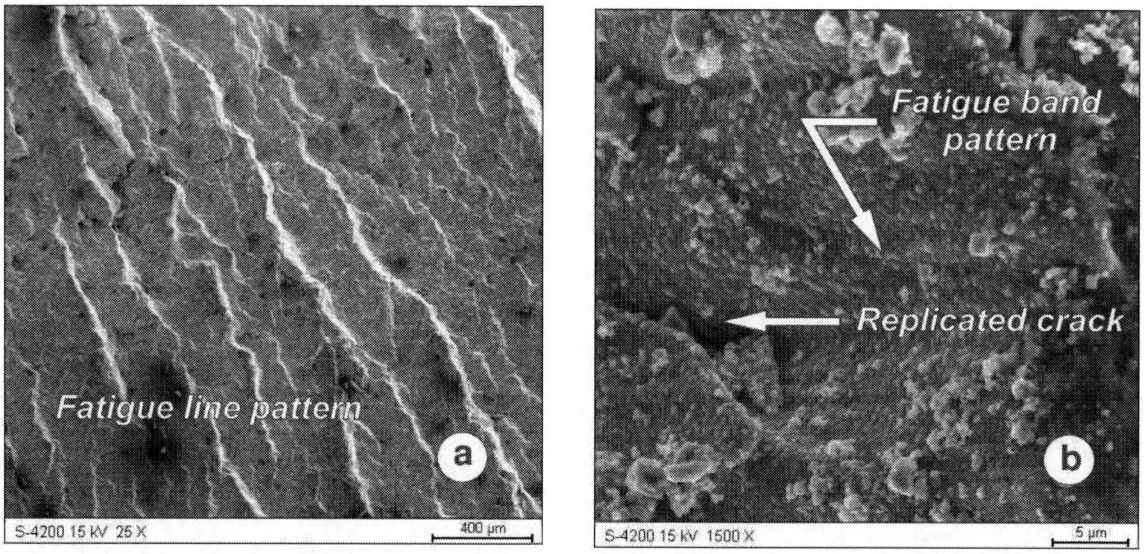

Figure 6. Fatigue scrap of spray tube („C" area on Fig. 3), (a) primary fault pattern running in direction of main crack propagation, (b) replicated cracks and fatigue bands

3.2. Microstructure of Spray Tube

Observations of polished sections in the light microscope disclosed the injector material in areas remote from outside surface of the spray tube has a bainite & ferrite structure typical for this steel grade (Fig. 7a). The effects of microstructure decomposition into ferrite along with carbide leaching and decarburization of the material were found on the outside surface of the spray tube (Fig. 7b).

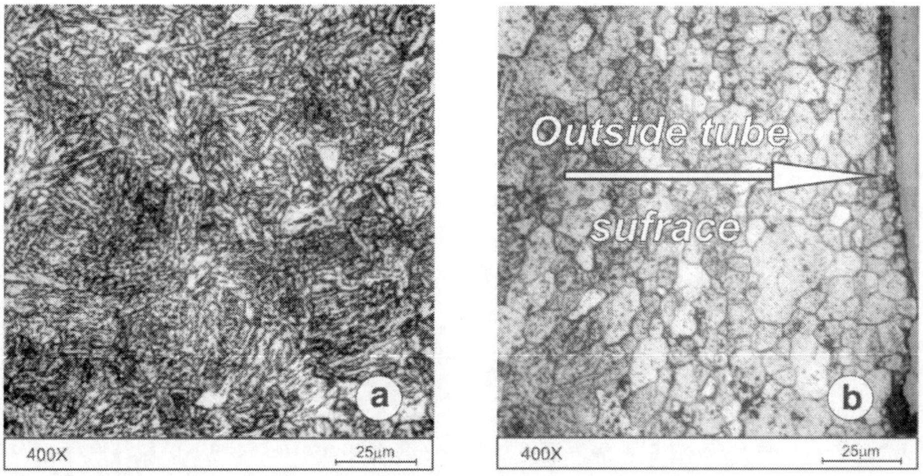

Figure 7. Microstructure of spray tube (a) Bainite & ferrite structure of deep section of tube material, (b) decarburization of outside tube surface

3.3 Mechanical tests

Fracture investigations disclosed that the spray tube cracks are of fatigue nature. The fracture particulars prove that cracking initiation and propagation was running perpendicularly to the injector centerline. (Fig. 3). It means the load which caused the cracking process was applied perpendicularly to the direction of steam flow through the chamber.

Generation of transverse forces acting on a cylindrical object located in a stream of flowing fluid is related with breaking-off an on-the-wall layer and forming an areo- & hydrodynamic wake - so called Karman vortex street as shown on Fig. 8 [8].

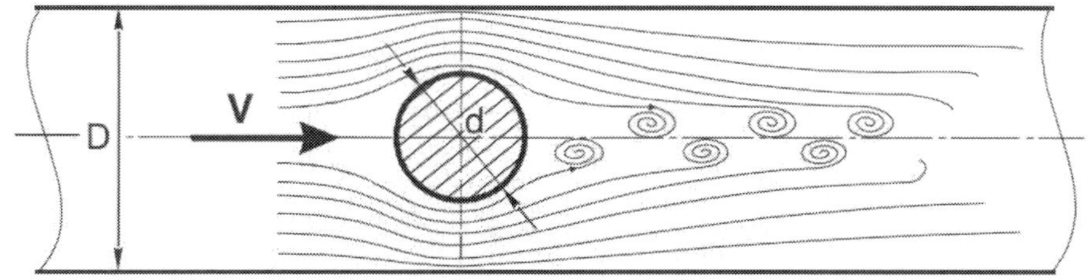

Figure 8. Formation of Karman vortex street

Flow of a fluid around cylindrical obstacle is perturbed by generation of regular vortices. These vortices, which flow down alternately to different side of the object generate harmonically varying force acting on the cylinder in the direction perpendicular to the direction of the fluid flow. Value of this force is calculated with the use of the following formula: [8]:

$$F_K = C_K \cdot \frac{1}{2}\rho \cdot v^2 \cdot S_d \cdot \sin(\omega t) \quad (1)$$

where: F_K - Karman vortex force [N]

C_K – dimensionless coefficient of Karman vortex force ($C_K=1$)

ρ - fluid density [kg/m^3]

v – velocity of steam flow through attemperator chamber [m/s]

Sd – plan view area of cylinder on which the vortices act [m^2]

ω – angular frequency of force action; $\omega = 2\pi f$

The relation between frequency of occurring actions, cylinder diameter and flow velocity is described by the following formula:

$$f = \frac{St \cdot d}{v} \tag{2}$$

where: f – Karman vortex frequency [Hz]
St– Strouhal number
d – cylinder diameter [m]

Velocity of steam flow through an attemperator chamber used in formula (1) may be calculated by means of the following formula:

$$v = \frac{\dot{m}}{S \cdot \rho} \tag{3}$$

where: $\dot{m}$ – mass-flow rate of steam through attemperator chamber [kg/s]
S – cross-section area of attemperator chamber with consideration of shock absorbing screen [m^2]
ρ – density of flowing steam [kg/m^3]

Generally, the Strouhal number for flow of considerably fluctuating velocity is not a constant value and depends on the Reynold number [9]. This relation $St=f(Re)$ is shown on Fig. 9.

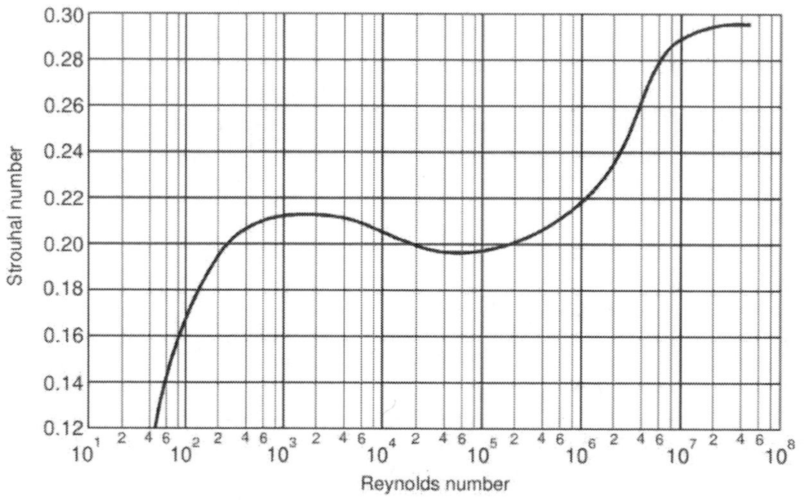

Figure 9. Strouhal number in function of Reynolds number

For determination of Strouhal number corresponding to the conditions of steam flow through attemperator chamber, the Reynolds number was calculated with use of the following formula:

$$\mathrm{Re} = \frac{\rho \cdot v \cdot D}{\mu} \tag{4}$$

where: v – average velocity of steam flow through attemperator chamber [m/s]
D – diameter of attemperator chamber with consideration of shock absorbing screen [m]
μ - dynamic viscosity [Pa·s],

For operating parameters of steam attemperator : p=20.6 MPa, t=520^{0}C, density and dynamic viscosity of steam flowing through attemperator chamber were calculated, and their values amount to $\rho = 66.72$ kg/m^3, and $\mu = 3.0748 \cdot 10^{-5}$ Pa·s, respectively [10]. Upon consideration of geometry parameters: D=0.2456 m, d=0.057 m, the velocity v=25.2 m/s and the Reynolds number Re=$1.35 \cdot 10^7$were calculated, and corresponding Strouhal number was read on Fig. 8.
Frequency of generated vortices calculated in the formula (2) is of f=129 Hz, and the force acting on injector chamber (Fig.1) is described by the following equation:

$$F_K \sim 180 \cdot \sin(2\pi \cdot f) \text{ [N]} \tag{5}$$

For evaluation of Karman force impact on magnitude of stress in a spray tube, a finite element method was used. In the model, the side surface of injector cylinder was loaded with a harmonically varying Karman force (5). Modal suppression with a factor of $\zeta = 0.05$ was assumed for calculations. As the pressure of steam and spraying water were similar, the impact of pressure of flowing media on magnitude of stress in the spray tube was disregarded. For modelling purposes, temperature-dependent, linear-elastic characteristics of a material was assumed. The results of calculations are shown on Fig. 10.

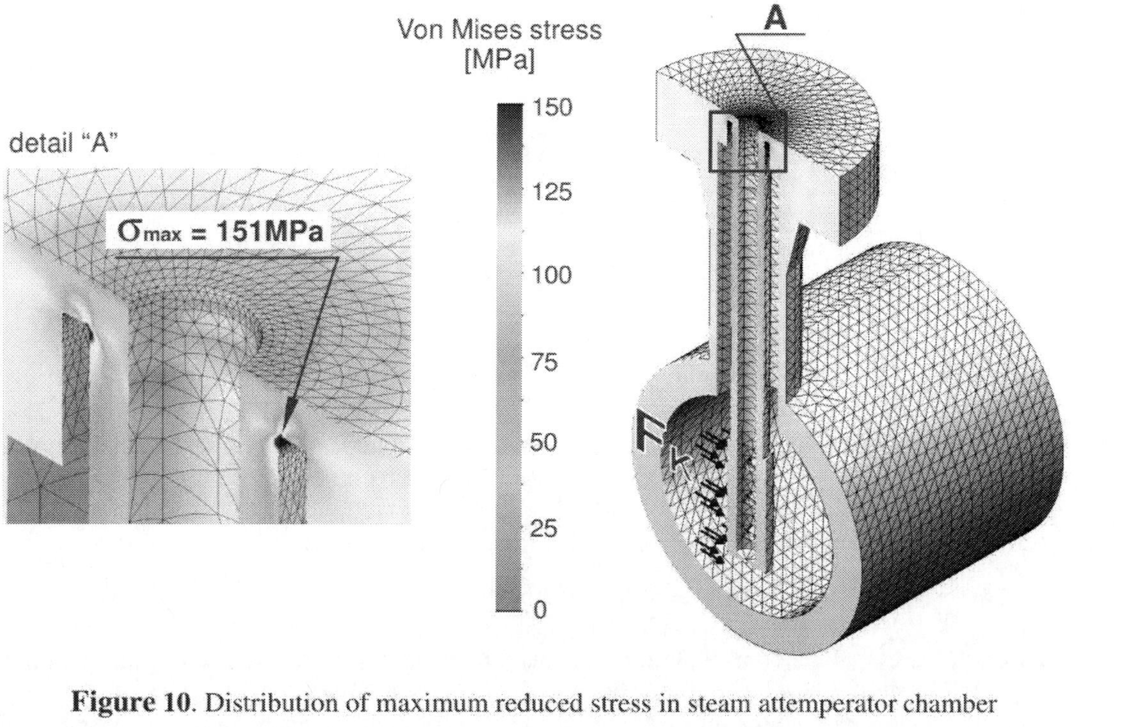

Figure 10. Distribution of maximum reduced stress in steam attemperator chamber

The results of calculation demonstrate that, as expected, maximum stress caused by harmonic load of Karman vortices occurs around the weld joining the spray tube with a flange. Calculated maximum stress value is a result of steam attemperator operation in conditions close to the resonance conditions. For the construction of steam attemperator shown on Fig. 1, the frequency of proper vibrations amounts to approx. 121Hz and is close to the frequency of the force exciting the system vibrations, which is of 129Hz, static load equal to Karman's force applied to the injector would cause a stress of approx. 39 MPa with consideration of geometry coefficient of tube -flange connection.
Considering the nature of spray tube cracking (Fig. 3, 4, 5), assumption may me made that the load action rarely occurring during operation of power boilers and disregarded by the designer was the main reason of damages occurred to the analysed element.

In the analysed steam attemperator chamber, additional action having impact on maximum value of stress in the spray tube is hydrodynamic reaction of spraying cooling water, value of which may be calculated according to the following formula [11]:

$$R = \frac{\dot{m}_1^2}{n^2 \cdot S_d \cdot \delta_1} \tag{6}$$

where: $\dot{m}_1$ - mass-flow rate of cooling water [kg/s]

n – number of spray nozzles

S_d – cross-section of spray nozzle [m^2]

δ_1 - density of water [kg/m^3]

Flow rate of water through the injector fluctuates considerably and is subject to boiler operation conditions. Changes of water mass-flow rate through injectors recorded within the selected portion of boiler operation period are shown on Fig. 11.

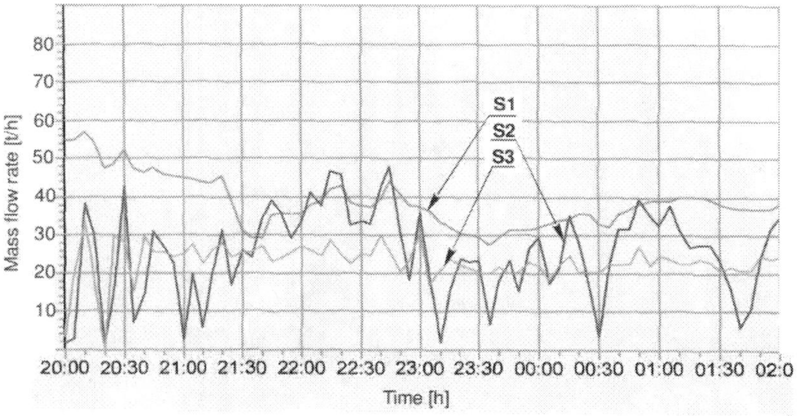

Figure 11. Display of changes of water flow rate through injectors of various types within the selected portion of boiler's steady operation time

Maximum value of hydrodynamic reaction of one nozzle calculated according to the formula (6) amounts to R=78 N. For calculation, maximum mass-flow rate recorded for S2 injector amounting to $\dot{m}$ =48 t/h and its geometry parameters shown on Fig.1 and 12 were used.

Spray tube bending moment caused by hydrodynamic reaction is equal to:

$$M_g = \sum_{i=1}^{3} R_i \cdot l_i = R \sum_{i=1}^{3} l_i \tag{7}$$

and causes the stress of random evolution and maximum value of 53 MPa with consideration of geometry coefficient of tube -flange connection.

The calculations made proved the major factor responsible for injector destruction was its operation in conditions of mechanical resonance. The effect of stress increase caused by resonance may be mitigated by change of frequency of proper injector vibrations at the set boiler operation conditions. The solution, which does not require extensive intervention in injector design and considerably improves its stiffness, is the use of a stiffener for permanent connection of injector chamber with steam attemperator chamber (Fig. 13).

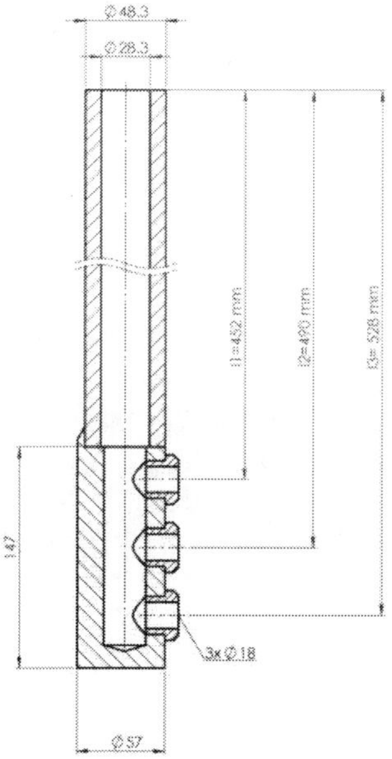

Figure 12. Overall dimensions of S2 injector.

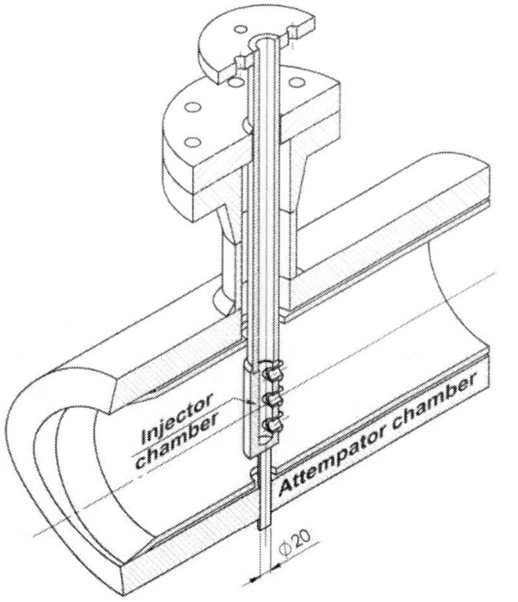

Figure 13. Modification of injector design

Calculation of stress distribution in modified steam attemperator chamber proved the maximum stress value was drastically reduced and does not exceed 10MPa (Fig. 14). Reduction of stress is a result of change in conditions of injector support and large difference between the frequency of exciting force and the frequency of injector's proper vibrations, which amounts to 495Hz.

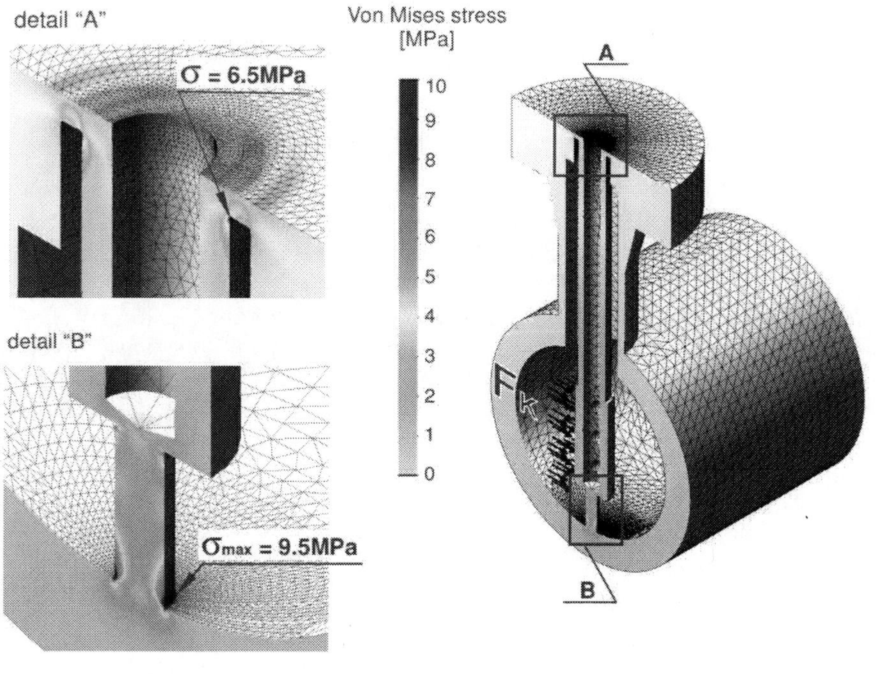

Figure 14. Distribution of maximum reduced stress in modified steam attemperator chamber

4. Summary

The results of fracture investigations and mechanical tests demonstrate the main destruction mechanism, which led to cracking of spray tube, was high-cycle fatigue of material occurring in conditions of oscillatory bending. Outside areas of the spray tube located outermost from attemperator chamber centerline were the centres of fatigue cracks. The reason for vibrations of spray tube, which caused stresses responsible for element destruction, was the effect of mechanical resonance caused by exciting force of frequency close to the frequency of proper vibrations of injector. The effect of resonance was caused by disregarding at the engineering stage of harmonically excited vibrations caused by vortices of flowing medium.

Observations of operating injectors, which are of similar design and operate in comparable conditions, but have different dimensions and different number of spray nozzles, confirmed the foregoing cause of cracking the analysed injector of S2 type. Fatigue cracks have not been found the destruction mechanism of the injectors of this type, and the most likely reason of their longer durability is a large difference between the frequency of proper vibrations and the frequency of exciting force.

For the analysed injector, calculated amplitudinal stress amounting to 150 MPa should not lead to spray tube cracking, because the fatigue limit Z_{rc} for steel grade 13CrMo4-5 in the attemperator's operating conditions is of approx. 350 MPa [12]. In the analysed case, the fatigue crack propagation was supported by surface decarburization of steel and direct adherence of HAZ, which formed additional structural notch. Furthermore, the area of welded joint of the spray tube with a flange formed a geometrical notch, which caused a stress concentration leading to initiation of fatigue cracking.

The manner by which the fatigue cracking of spray tube will be eliminated is a re-designing aiming at changing the frequency of proper vibrations of the element. Implementation of solution shown on Fig. 14 will cause considerable reduction of stresses in the analysed element by 15 times and requires only minimum modification of steam attemperator design.

The stress caused by steam flow will not exceed 10MPa and will have negligible effect on life of modified steam attemperator chamber.

5. References

[1] Hernas A, Dobrzański J 2003 *Durability and Destruction of Boiler and Turbine Components* (Publishing House of the Silesian Technical University, Gliwice)

[2] Mutwil K, Cieśla M 2007 Constructional and Operational Conditions of Steam Superheater Cracking *Materials Engineering*. **6** 914

[3] Okrajni J, Mutwil K, Cieśla M 2007 Steam Pipelines Effort and Durability *Journal of Achievements in Materials and Manufacturing Engineering* vol. **22,** issues **2** 63

[4] Mutwil K, Cieśla M, 2004 Assessment of Operation Efforts of Steam Superheaters in Steady Conditions of Power Boiler Operation *Computer Aided Engineering* **9** 693

[5] Mutwil K, Cieśla M, Okrajni J 2008 Predicting the Life of Power Generation Facilities with Use of Numeric Modelling *Gloryhole Mining* **4-5** 141

[6] Kocańda S 1985 *A fatigue cracking of metals* (WNT, Warszawa)

[7] Renowicz D, Cieśla M 2007 Crack initiation in steel parts working in boilers and steam pipelines *Journal of Achievements in Materials and Manufacturing Engineering* vol. **21**, issue **2** 49

[8] Den Hertog J 1978 Mechanical Vibration (WNT, Warszawa)

[9] Drobniak St 1997 *Fluid Mechanics Laboratory* (Publishing House of the Częstochowa Technical University, Częstochowa)

[10] Raznievic K 1966 *Thermodynamic Values with Graphs* (WNT Warszawa)

[11] Miler A 1998 *Heat and Power Engineering Machines and Equipment* (WSP, Warszawa),

[12] Collective work 1978 *Steel Specification. Steels for Application at Elevated and Low Temperature* (Publishing House Silesia, Katowice)

IOP Conf. Series: Materials Science and Engineering **35** (2012) 012018 doi:10.1088/1757-899X/35/1/012018

Preparation of translucent YAG:Ce ceramics by reaction sintering

D Michalik, M Sopicka-Lizer, T Pawlik, J Chrapoński
Department of Materials Science, Silesian University of Technology, Krasińskiego 8. 40-019 Katowice, Poland

E-mail: daniel.michalik@polsl.pl

Abstract. The paper presents preparation of YAG precursor powders mixture with ability for the reaction sintering at lower temperature. The powder mixture was mechanochemically processed and characterized for the specific surface area and crystallinity (XRD). Sintering experiments were performed on the batched powders, mechanochemically processed powders and as-synthesised YAG powders. XRD measurements and SEM/TEM studies were applied for characterization of the phase composition and development of YAG ceramics. Transmittance of the resultant ceramics was measured. It has been shown that YAG structure formed at 200 °C lower temperature in comparison to the reference specimen and contamination by the side phases (YAM, YAP) was limited. The best translucency was achieved for the reaction sintering of the mechanochemically processed powders.

1. Introduction

Yttrium Aluminum Garnet (YAG) is a well known host lattice for luminescent materials. Presently, YAG in the form of monocrystals is applied in diode-pumped solid-state lasers, scintillators, IR windows [1]. The development of blue light emitting diode (LED) has opened a way to obtain white light for general lighting by using phosphors for partial conversion of blue light into the yellow one. The optical mixture of the blue and yellow light results in a white light. Commercial white LEDs (wLEDs) employ mixtures of phosphors powders deposited on a GaN semiconductor chip and YAG powder doped by Ce^{3+} was the first and an efficient blue light converter because of 4f-5d transition and maximum absorbance in the blue region [2].

Thin transparent YAG ceramics with the relevant doping could replace both the monocrystals or the resin-based YAG:Ce composites as the blue light converters in wLEDs [3]. The advantages of the monocrystal replacement by polycrystalline transparent ceramics are obvious: low price, ease of manufacturing and mass-production [4]. It also has been proven that the solubility of Nd for polycrystalline ceramics is more than that of a single crystal [5]. Substitution of the resin-based phosphor converter in wLEDs by transparent polycrystalline ceramics will result in the significant decrease of operating temperature because thermal conductivity of YAG ceramics is approximately 40 times higher in comparison to the resin-based composites. Consequently, thermal quenching of YAG:Ce phosphor will be drastically reduced and efficiency of wLED will be considerably improved.

Recently, a lot of efforts have been made for manufacturing transparent polycrystalline YAG ceramics with various optical activators [6-8]. The cubic structure of garnet enables fabrication of the polycrystalline YAG ceramics with high transparency [9].
The YAG powders may be prepared and synthesized by different methods. The preparation method of YAG nanopowders includes: sol–gel [10-11], coprecipitation [12], hydrothermal [13], spray pyrolysis [14], combustion [15], solvothermal methods [16]. The other method of preparation nanopowders is mechanochemical processing (MCP) as a additional step in the solid state reaction method. High energy ball milling initiate chemical reactions and structural changes by transferring kinetic energy to the grain refinement of milling mixture [17]. However, placing of optical activators into the YAG precursor by MPC and subsequent sintering was never reported before.
The resultant nanopowders after necessary calcination can be then consolidated into polycrystalline ceramics by one of the following methods: pressureless sintering at temperature over 1700 °C followed by densification in high vacuum [9] or pressure assisted methods i.e. hot-isostatic pressing (HIP) in the range of 1450-1550 °C [18], spark plasma sintering (SPS) at 1300 °C [19] or low temperature high pressure method (LTHP) [20]. There are only few reports on manufacturing transparent YAG:Ce ceramics [20-22] as cerium ions were found to segregate at grain boundaries as revealed lately [23].
The aim of this study was to obtain transparent YAG:Ce ceramics by a one-step reaction sintering using a solid state reaction method for densification of mechanochemically processed mixture of powdered substrates. It was expected that the activation of powder will amplify sintering capability by the resultant increase of the powder surface area [24].

2. Experimental

The stoichiometric amount of the high purity oxides according to the formula of $Y_{2.94}Ce_{0.06}Al_5O_{12}$ were batched.η-Al_2O_3 (Sasol, 99.99 %), Y_2O_3 (Stark, 99.99 %) and CeO_2 (AtomicChemetals, 99.99 %) powders with anhydrous ethanol (99.8 %) were ultrasonically deagglomerated and mixed in a roller bench in polypropylene bottle with 10 mm Si_3N_4 balls for 24 hours. The suspension was dried at room temperature for 24-48 hours and grinded in an agate mortar. The mechanochemical process was conducted by Fritsch Premiumline 7 planetary ball milling system. The parameters of high-energy milling were: rotation speed of 1000 rpm for 15 min, followed by a break of 30 min for cooling milling jars. Such a cycle was repeated 8 times. The total time of milling was 2 hours. A 80 ml silicon nitride milling jar and 5 mm silicon nitride ball were used. A weight ratio between the milling balls and powder was 15:1. Dilatometer studies were performed on the thermal analyzer SETSYS using TMA head in an argon atmosphere (Ar 99.999%) with a flow of 1.25 l/h. The heating rate was 5 °C/min, and the maximum temperature was 1500 °C. The linear changes of both the as-batched powder and the milled precursor were examined.
Three ways of specimen consolidation were applied. The first set of specimens was prepared in a two-step procedure. First, mechanochemically processed precursor powder was annealed for 1 hour in an N_2/CO atmosphere in a graphite furnace (Thermal Technology) at temperature of 1650 °C. It was checked for phase composition and consequently it was used for consolidation at 1650 °C in the form of pellets (diameter of 10 mm, height of 3 mm) -sintering of YAG:Ce powder (2s-S). The second type of specimens were compacted directly from MCP-ed powders and then sintered in a graphite furnace at 1650 °C for 2 hours (reaction sintering of the activated precursor –RS). The last type of specimens was prepared as reference specimens. They were uni-axially pressed from the as-batched mixture of the relevant oxides without MCP-step and consequently consolidated at 1650 °C for 2 hours (reaction sintering of initial powders – reference RS). Diameter and height of all compacted specimens were the same. All of them were sintered in nitrogen flow of 3 l/h in a graphite furnace on an alumina support.
Phase composition of the powders and the final ceramics was examined by X-Ray powder diffraction (XRD) using a Philips XPert PRO MPD system equipped with linear detector X'Celerator. Lattice parameters were obtained from Rietvield refinement of the XRPD data, recorded in the 2Θ range of 10°-80° with a step size of 0.02°. The grain morphologies of the powders and surface of the resultant

ceramic were observed in a Hitachi S-3400N scanning electron microscope (SEM) operated at 15kV. All measurements were performed at room temperature.

3. Results and discussion

The surface area of starting powder and MCP-ed powder were presented elsewhere [23]. The results were 25.6 m^2 g^{-1} and 42.4 m^2 g^{-1}, respectively. Dilatometry studies allow watching the specimen's height changes during thermal treatment. Since several reactions are possible in the Al_2O_3-Y_2O_3-CeO_2 system the subsequent height changes are related to the reactions between the components and subsequent volume changes due to the new phase crystallization. Thermal behavior of alumina, yttria and ceria mixture during increasing temperature in argon atmosphere is given in Fig.1. It shows clear difference of the samples thermal behavior if the MCP step was introduced into powder processing procedure.

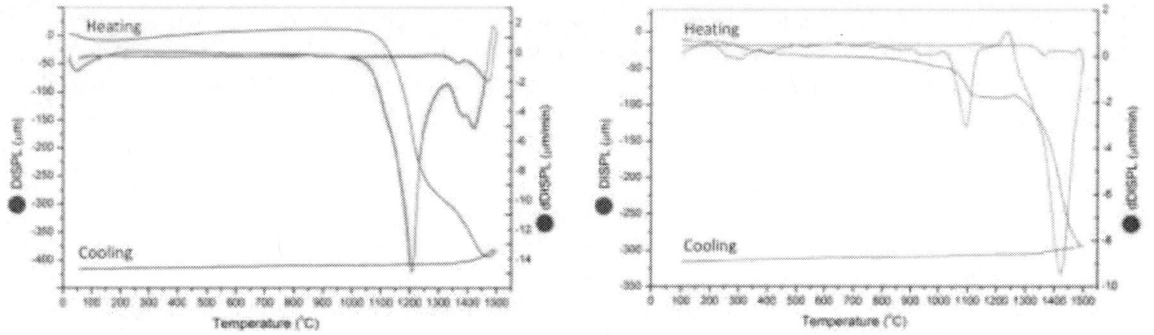

Figure 1. Dilatometer study of the YAG:Ce powders prepared by the reaction sintering method: without preliminary MCP (a) and with the MCP step (b).

If the specimen was compacted from the as-batched mixture then the beginning of the reaction at 980 °C was noted (Fig.1a). The highest rate of the first reaction with significant volume changes occurred at 1200 °C and according to literature [25] it could be ascribed to yttrium aluminum monoclinic (YAM) - phase formation. This reaction is followed by the next two ones with smaller volume changes related to the formation of yttrium aluminum perovskite (YAP) and finally to YAG phases, however complexity of the peaks hinders the proper solution of the resultant peak. Behavior of the sample with the MCP step (Fig.1b) allows to detect the occurrence of three reactions, two of them related to the volume decrease and one associated to the increase of volume. Comparison of YAM and YAP densities evidences the volume enlargement after transition-reaction completion and that performance is clearly supported by the recorded thermal analysis. The last reaction with the highest volume changes could be related to YAG formation assisted by sintering. Moreover, if we compare behavior of both powders then a shift of the peaks toward the lower temperature is observed: crystallization of YAM and YAP phases occurs at temperature by 100 °C lower in comparison to the unmilled precursor. Moreover, the largest displacement of the milled specimen occurred at the highest temperature and was related to YAG crystallization and simultaneous consolidation. In contrary, the unmilled precursor demonstrated the largest displacement during crystallization of YAM while densification at 1420 °C involved onlythe limited shrinkage of the specimen.

It should be noted that dilatometer curves during cooling show small volume changes. Those changes occur in the both tested specimens and at the same temperature. Since the phase composition of the both heat treated specimens is obviously different we think that the observed transition could be related to the cerium presence.

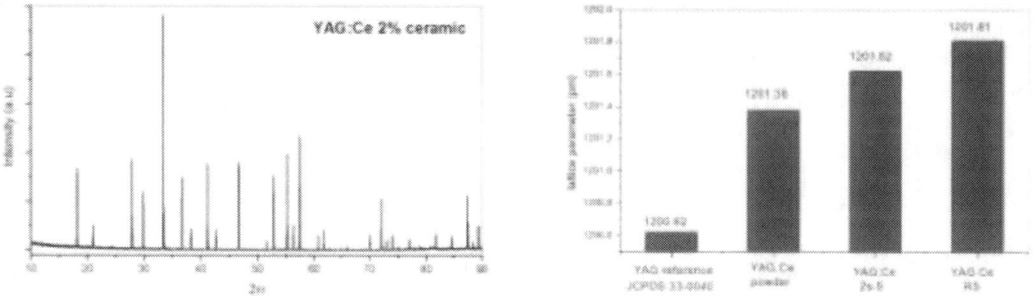

Figure 2. Typical XRD patterns of YAG:Ce powder and ceramics (a) and the lattice parameter of YAG:Ce powder/ceramics after processing/sintering (b).

Fig 2a presents the typical pattern of YAG:Ce powders and ceramics obtained during the experiments. All samples showed the pure garnet structure with high level of crystallization [26]. Figure 2b shows the changes of the YAG:Ce lattice parameter at a diverse step of processing. The lattice parameter of the pure YAG is 1200.62 pm. The lattice parameter goes up as a result of substitution of Y^{3+} (102 pm) by Ce^{3+} (115 pm) ions. It was observed that the YAG: Ce powder after synthesis has a lower lattice parameter than YAG:Ce ceramics. The two-step sintered (2s-S) YAG:Ce ceramic shows lattice parameter of 1201.62 pm, higher than the powder before sintering. Increase of the lattice parameter after sintering must be related to the higher substitution of yttrium ions in the garnet structure by Ce^{3+}. It could mean that only part of cerium ions entered the garnet structure during the solid state synthesis leaving the remaining cerium oxide on the surface of garnet particles. It could occur since particles of the three different powders were only loosely packed. Moreover, only Ce^{3+} ions could enter the garnet structure and necessary reduction of Ce^{4+} involves growth of cerium ion from 101 pm (Ce^{4+}) to 115 pm (Ce^{3+}). The subsequent densification of the partially substituted garnet powder allowed completion of the designed solid solution as a result of shortening the diffusion paths in the compacted specimens. However, if the mechanochemically processed powder is directly reaction sintered then the lattice parameter is the highest. This suggests that the reaction sintering of the MCP-ed mixture guarantees better conditions for the reduction of Ce^{3+} ions and subsequent formation of the solid solution. Such behavior was observed for the many reactive mixtures [27] however it was never reported for cerium doped YAG-based materials. SEM images (Fig.3) of the initial and activated powders show the results of high-energy milling.

a) as-batched powders b) powders after MCP

Figure 3. SEM images of the initial powders (a) and powders after MCP (b).

IOP Conf. Series: Materials Science and Engineering **35** (2012) 012018 doi:10.1088/1757-899X/35/1/012018

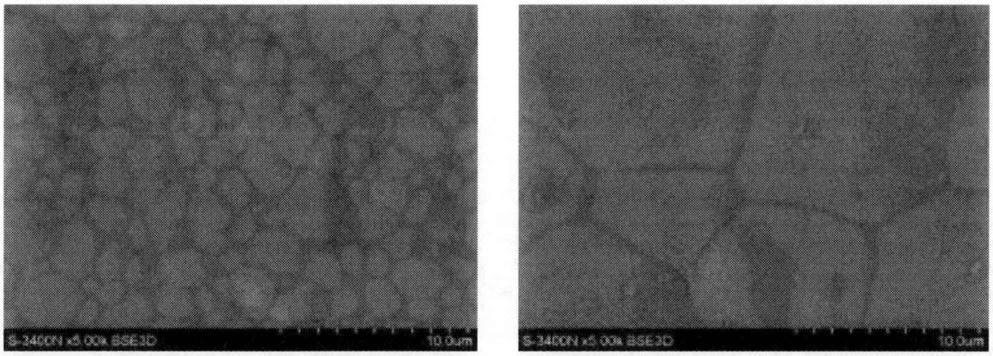

Figure 4. SEM images of surface of ceramics samples by the two-step sintering - 2s-S: (a) and reaction sintering - RS (b) at 1650 °C.

Figure 4 shows the SEM images of the YAG:Ce ceramics surface. The significantly larger grain growth was obtained for RS samples with the MCP step. Change of the grain sizes from 0,5-3 μm to 5-10 μm was observed. The latter confirms high sintering ability of the MCP-ed powders and opens possibility for manufacturing transparent YAG:Ce ceramics. The YAG:Ce disks of 0.3 mm thickness were translucent and 20% light transmittance was observed for the polished samples.

4. Conclusions

The mechanochemical processing of the stoichiometric mixture of oxide powders relevant to the formula of $Y_{2.94}Ce_{0.06}Al_5O_{12}$ resulted in significant changes of the reaction sequence and sintering ability during thermal treatment: formation of the transient phases of YAM and YAP was moved to temperature lower by 100 °C while crystallization of YAG phase was followed by the substantial shrinkage related to consolidation Consequently, it has been shown that translucent YAG:Ce ceramics could be obtained by pressureles one-step reaction sintering accompanied by the extensive grain growth. Moreover, the MCP step increased solubility of Ce^{3+} ions in the garnet structure as evidenced by a change of lattice parameter from 1201.38 pm (powder) to 1201.81 pm (ceramic). Consequently, the significant reduction of the necessary concentration of that optical activator in the initial powders could be achieved.

Acknowledgement

The financial support by Polish Ministry of High Education under the project No: N N507 467737 is gratefully acknowledged.

References

1. Zorenko Y, Voznyak T, Gorbenko V, Zych E, Nizankovski S, Danko A and Puzikov V 2011 *J. Lumin.* **131** 17.
2. Nakamura S, Senob M, Iwasa N and Nagahama S 1995 *Jpn. J. Appl. Phys.* **34** 797.
3. Chao W-H, Wu R-J, and Wu T-B 2010 *J. Alloys Compd.* **506** 98.
4. Wen L, Sun X, Xiu Z, Chen S, and Tsai C-T 2004 *J. Eur. Ceram. Soc.* **24** 2681.
5. Ikesue A 2002 *Opt. Mater.* **19** 183.
6. Nishiura S, Tanabe S, Fujioka K, Fujimoto Y and Nakatsuka M 2009 *IOP Conf. Series: Materials Science and Engineering 1* doi:10.1088/1757-8981/1/1/012031.
7. Pokhrel M, Kumar G A, Samuel P, Ueda K I, Yanagitani T, Yagi H and Sardar D K 2011 *Opt. Mater. Ex.* **1** 1272.
8. Zhang W-X, Pan J-B, Zhou J, Liu W-B, Li J, Zou Y-W and Wei Z-Y 2011 *Ceram. Inter.* **37** 1133.
9. Nishiura S, Tanabe S, Fujioka K and Fujimoto Y 2011 *Opt. Mater.* **33** 688.
10. Veith M, Mathur S, Kareiva A, Jilavi M, Zimmer M and Huch V 1999 *J. Mater. Chem. Phys.* **9** 3069.

11. Xia G, Zhou S, Zhang J and Xu J 2005 *J. Crystal Growth* **279** 357.
12. Wang Y, Zhang L, Fan Y, Luo J, McCready D E, Wang C and An L 2005 *J. Am. Ceram. Soc.* **88** 284.
13. Takamori T and Dawid L D 1986 *Am. Ceram. Soc. Bull.* **68** 1282.
14. Kang Y C, Lenggoro I W, Park S B and Okuyama K 2000 *Mater. Res. Bull.* **35** 789.
15. Fernandez A and Somers J 2003 *J. Mater. Sci.* **38** 2331.
16. Li X, Liu H, Wang J, Cui H and Han F 2004 *Mater. Res. Bull.* **39** 1923.
17. McCormick P G and Froes F H 1998 *J. Met.* **50** 61.
17. Lee H, Mah T and Partasarathy T 2008 *Ceram. Eng. Sci. Proc.* **25** 147.
18. Frage N, Kalabukhov S, Sverdlov N, Ezersky V and Dariel M P 2010 *J. Eur. Ceram. Soc.* **30** 3331.
19. Pazik R, Głuchowski P, Hreniak D, Strek W, Ros M, Fedyk R and Łojkowski W 2008 *Opt. Mater.* **30** 714.
20. Li J, Chen Q, Feng G, Wua W, Xiao D and Zhu J 2012 *Ceram. Inter.* **38** 649.
21. Chang-qing L, Hong-Bo Z, Ming-fu Z, Jie-cai H and Song-he M 2007 *Trans. Nonferrous Met. Soc. China* **17** 148.
22. Zhao W, Anghel S, Mancini C, Amans D, Boulon G, Epicer T, Shi Y, Feng X Q, Pan Y B, Chani V and Yoshikawa A 2011 *Opt. Mater.* **33** 684.
23. Michalik D, Sopicka-Lizer M, Plewa J and Pawlik T 2011 *Arch. Metall. Mater.* **56** 1257.
24. Ikesue A, Kinoshita T, Kamata K and Yoshida K. 1995 *J. Am. Ceram. Soc.* **78** 1033.
25. JCPDS Database no. 33-0040
26. Sopicka-Lizer M 2010 *High-energy ball milling. Mechanochemical processing of nanopowders.* Woodhead Publishing Limited.

IOP Conf. Series: Materials Science and Engineering **35** (2012) 012019 doi:10.1088/1757-899X/35/1/012019

The effect of pressureless densification on mechanical and tribological properties of fine-grained silicon nitride ceramics

T Pawlik[1], M Sopicka-Lizer[1], J Wieczorek[2]
[1]Department of Materials Science, Silesian University of Technology, Krasińskiego 8. 40-019 Katowice, Poland[1]
[2]Department of Materials Technology, Silesian University of Technology, Krasińskiego 8. 40-019 Katowice, Poland

E-mail: tomasz.pawlik@polsl.pl

Abstract. The paper presents a new economic method of silicon nitride ceramic preparation for high wear resistant applications as sealing valves or the brake pad lining. The excellent mechanical properties and wear resistance of the resultant ceramic were improved by reduction of silicon nitride grain size to the one-two micrometer level as a result of mechanochemical processing and special procedure of compact densification. All experiments were conducted on specimens prepared from α-Si_3N_4-AlN-Y_2O_3 powders with application of mechanochemical processing (MCP). The chosen specimens were tested for hardness, elastic modulus, bending strength and wear resistance. The results showed mechanical properties in the range of hot-pressed ceramics and superior wear resistance due to micrometer-sized β-Si_3N_4 grains.

1. Introduction

Silicon nitride is recognised as one of the best candidates among the advanced ceramics for use in tribological application as valves, bearings, seal rings, cutting tools due to high hardness, high thermal conductivity, wear resistance and chemical inertness in the contact with metal. Accordingly, an extensive research has been conducted toward improvement their wear performance against friction and wear by the mean of composites preparation [1-2]. Recently, there has been significant effort to improve mechanical properties of polycrystalline ceramics by refinement of their grain size [3-4]. However, silicon nitride-based ceramics owe their excellent mechanical strength and fracture toughness to the elongated β-Si_3N_4 or β-sialon grains after densification by the means of hot pressing (HP) or gas pressure sintering (GPS). Refinement of the silicon nitride microstructure below one micrometer could be achieved by application of nano-sized or amorphous silicon nitride powder [5-6] followed by a rapid consolidation such as spark plasma sintering (SPS) [7-8] or two-step sintering at relatively low temperature [9]. The key factor in acquiring silicon nitride grains with a diameter below 1 μm is to suppress the grain growth by application fast densification (i.e. SPS) or by consolidation conducted at relatively low temperature. Application of mechanochemical processing (MCP) of the initial powders [10-11] accompanied by thorough homogenization of silicon nitride powder and sintering additives [12-13] can be a way for increasing sinterability of silicon nitride powders and reducing sintering temperature. That procedure involves high energy milling in a planetary mill in an appropriate gaseous environment in order to protect non-oxide powder particles against uncontrolled oxidations. The resultant powder is characterized by high compaction ability and capability for consolidation at relatively low temperature [14]. Possibility of pressureles consolidation opens the way for manufacturing a large variety of shapes and sizes of fine-grained monolithic silicon nitride ceramics. Since the importance of microstructure on the mechanical properties of silicon nitride and sialon ceramics is well known and documented [15-16], the newly developed fine-grained ceramics must be characterized and compared with the other examples of sub-micro or nanoceramics in terms of their mechanical and tribological behaviour. In this work, we report on the effect of sintering parameters of the mechanochemically activated Si_3N_4-based precursor on the mechanical and tribological properties of the resultant fine-grained silicon nitride ceramics.

2. Experimental

2.1. Sample preparation

A mixture of commercial powders of α Si_3N_4 (M11 grade by H. S Starck) and additives as: 5.7 wt% of high purity AlN (fine C grade by H. S Starck) and 5 wt% of Y_2O_3 (fine C grade by H. S Starck) powders were chosen. Aluminium nitride has been chosen as an additive since some oxidation of nitride particles during MCP could have occurred. The mixture of methylethyleneketone (MEK) (butanone-Sigma Aldrich) and ethanol (ET, 99.8 vol.% by POCH) in 66/34 %vol ratio was used as a solvent for the preparation of a suspension. Hypermer KD1 (3 wt%) was added to MEK/ET suspension as a surfactant. Homogenization process was a multi stage one and it consisted of the following steps: mechanical mixing of the suspension on a roller bench for 24-48 hours, next ultrasonical treatment for 2 hrs (Branson Sonifier W-450D) and finally drying in a laboratory dryer at 110 °C. Dried powder was then milled (MCP) in a high energy micro-planetary mill (Pulverisette Premium Line 7 by Fritsch). The vial with Si_3N_4 lining and silicon nitride grinding balls of 5 mm diameter were used in a high purity argon atmosphere. MCP was performed at a rotation speed of 1000 rpm for 0.5 hours. Ball to powder ratio was 4:1. The resultant powder was then compacted into pellets of 3-5 mm height and 10 or 30 mm diameter by preliminary uniaxial pressing at 50 MPa followed by cold isostatic pressing (CIP) at 300 MPa. The resultant pellets were thermally treated in air (500 °C for 2 hrs) in order to decompose and to remove organic additives. Subsequently, samples were pressurelessly sintered in a graphite chamber furnace (Thermal Technology) in a powder-bed (BN/Si_3N_4) at temperature of 1500 °C in high purity (99.999%) nitrogen flow for 240 or 360 min. In order to improve sintering ability without grain growth the consolidation process was modified by initiation of densification at temperature of 1700 °C (1-2 minutes) followed by rapid cooling (furnace cooling) to 1500 °C in order to reduce grain boundary mobility and soaked at this temperature for 240 or 360 min.

2.2 Sample characterization

The apparent density of the sintered specimens was measured using the Archimedes method in water. Theoretical density was calculated from the mixture rule for density of β-Si_3N_4, α-Al_2O_3 and Y_2O_3. X-ray diffraction (XRD) (Cu Kα radiation) was performed on a polished cross-section of the specimens. Hardness was measured by Vickers indentation method in a HV10 test (load of 98 N) in Struers Duramin A 300 equipment. Elastic modulus was obtained from the ultrasonic wave velocity measurement. Fracture toughness K_{IC} was estimated from the cracks length measurement based on Anstis's formula after indenting at the same load as for hardness test for 15 s [15]. 5-6 specimens with a size of 1x3x20 mm were prepared by cutting the pellets followed by machining and polishing the specimen surface. Bending test was carried out using a three-point loading device with a span of 10 mm and a cross-head speed of 2 mm min^{-1} at room temperature.

Erosion tests were performed on the selected samples with a polished surface by the mean of impacting Al_2O_3 abrasives powder (particles of average diameter of 50 μm) in the air flow. 45° angle of attacking powder beam was used in order to avoid an adhesive deposition of abrasives grain on the specimen surface. Medium (30 m s^{-1}) to high (70 used in order to avoid an adhesive deposition of abrasives grain on the specimen surface. Medium (30 m s^{-1}) rates of abrasives flow were applied for 10 min.

The resultant mass losses of the specimen were recorded. The wear behaviour was studied on the polished specimen's surface in ambient conditions (20±5 °C and 45%±RH) in the unlubricated pin on block experiment. The spheroid cast iron pin of 3 mm diameter was used. The sliding tests were performed at various number of cycles: 20 000 (distance of 240 m), 40 000 (distance of 480 m) and 58 000 (distance of 696 m) with application of 18 or 36 N load and constants sliding speed [0.05 m s^{-1}]. Mass losses of both, the specimen and the pin, were recorded. Wear rate k was calculated according to the formula: $k = \frac{W}{F_N * s} < 10^{-6}\left[mm^3 N^{-1} m^{-1}\right]$ (1) [16],

where W is the wear volume of specimen (mm^3), k is the wear rate in (mm^3 N^{-1} m^{-1}), P is the normal load (N), s is the sliding distance (m). The track width was measured in the light microscope while the surface profiles of the worn surface were acquired using profilometer SRT microProf 3000 to measure the depth of the wear tracks. During the sliding tests, the frictional forces were recorded continuously during the testing, from which the friction coefficient was calculated. Fractured and worn surfaces of the materials were observed in the Scanning Electron Microscope (Hitachi 3400N).

3. Results and discussion

All investigated materials were of high density (calculated density is close to 3.29 g/cm^3) with mono-phase composition of β-sialon with a low z-value of 0.2. Characterization of the resultant samples is given in Table 1 and Fig.1. Low-substituted β-sialon formed as a result of nitrogen excess in the initial mixture since oxidation of AlN during MCP has not been completed.

Table 1. Properties of silicon nitride ceramics sintered at different temperature regimes

Temperature of sintering (°C)	Sintering time (min)	Apparent density (g cm^{-3})	Hardness (MPa)	K_{IC} (MPa m$^{0.5}$)	Elastic modulus (GPa)	Wear rate (mm^3 N^{-1} m^{-1})
1500	240	3.21	17.6±0.6	6.38±1.04	314	8.30*10^{-7}
1500	360	3.22	17.7±0.5	6.05±0.66	309	8.28*10^{-7}
1700$^{\#}$/1500	240	3.23	17.5±0.2	6.17±0.48	313	8.26*10^{-7}
1700$^{\#}$/1500	360	3.24	17.6±0.2	5.52±0.71	314	8.26*10^{-7}

$^{\#}$*fast heating to 1700 °C for 1-2 minutes, furnace cooling to 1500 °C and soaked for 240 or 360 minutes.*

The final density of the resultant specimens is very high even though the materials were sintered at low temperature of 1500 °C. Moreover, it is comparable with the density of similar initial composition derived silicon nitride ceramics (3.23 g cm^{-3}) densified by spark plasma sintering (SPS) at 1585 °C under pressure of 50 MPa [1]. The excellent consolidation of the resultant ceramics is confirmed by very high elastic modulus. Elongation of the sintering time slightly improves densification effects only if the two-step procedure was applied. Correspondingly, increase of temperature at the initial stage of sintering to 1700 °C and subsequent immediate drop to 1500 °C seems to improve densification effects as slightly higher density was observed in those specimens and the results are confirmed by the elastic modulus outcome.

The obtained silicon nitride ceramics show high hardness at the narrow range of 17.5-17.7 GPa while scattering of fracture toughness is higher than reported elsewhere [1-2] but the lowest value does not drop below 5.5 MPa m$^{0.5}$. In contrary to density results, two-step sintering as well as elongation of sintering time, they both slightly reduce fracture toughness. It was reported previously that silicon nitride ceramics with micrometer/submicrometer grain size had a modulus of 250-300 GPa, a hardness of 14-16 GPa and a fracture toughness of 3-6 MPa m$^{0.5}$ [17-18]. Standard hot pressed at 1850 °C 485-Si_3N_4 ceramics is reported to show hardness of 15.07 ±0.23 GPa and K_{1c} of 5.8 MPa m$^{0.5}$ [19]. Recently reported, Si_2N_2O secondary phase-free dense silicon nitride ceramics obtained by SPS is characterized by low hardness of 12.8-13.5 GPa depending on sintering temperature and the resultant grain size [20]. Hardness values over 18 GPa were reported only for specially designed α- or α/βsialons prepared by SPS or GPS (Gas Pressure Sintering) [19]. Therefore, comparison of the newly developed pressureless sintered ceramics with those manufactured by the advanced sintering techniques confirms their first-rate properties.

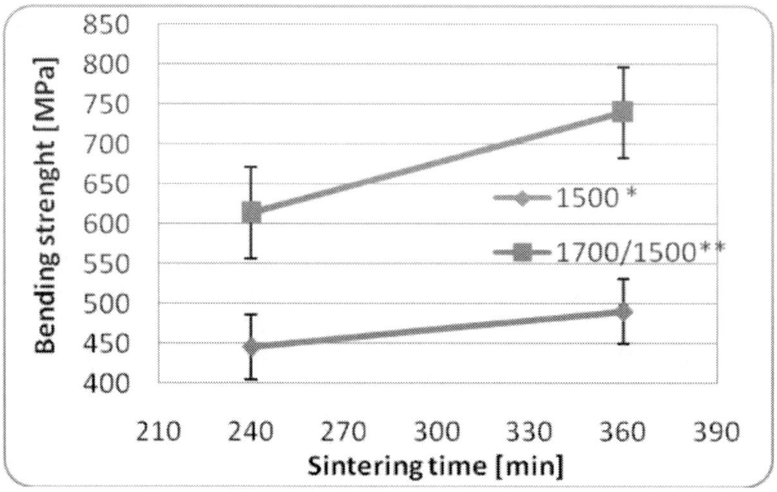

Figure 1. Influence of the sintering time and sintering way on bending strength. The bottom line – sintering at 1500 °C, *- one step reaction sintering, the upper line – two-step regime 1700/1500 °C, **- heating to 1700 °C for 1-2 minutes, furnace cooling to 1500 °C and soaked for 240 or 360 minutes)

The influence of sintering procedure on the bending strength of the tested specimens is shown in Figure 1. It is evident that two-step sintering improves mechanical strength of the material. Such behaviour must be related to the changes of the microstructure. Examination of the fractured surface in the SEM revealed various thickness and elongation of β-Si_3N_4 grains. Although no quantitative studies were performed, it could have been noticed that larger number of elongated grains was present in the specimens produced by the method of two-step sintering (Fig. 2). It is supposed that rapid cooling from 1700 °C to 1500 °C must have resulted in spontaneous nucleation of numerous β-Si_3N_4 nuclei along with decreasing solubility. In contrary, precipitation of β-Si_3N_4 at 1500 °C was slower. Apart from that, absence of external pressure during sintering enhanced grain growth.

Figure 2. Microstructure (SEM) of the fractured sample produced by the two-step sintering regime (1700/1500 °C) by 360 min.

The results of erosion tests showed the excellent resistance of the material under angle attack of 45° as very low mass losses in a narrow range of 2.5 to $8.9*10^{-3}$ wt% were recorded for all tested specimens. However, microscopic examination of the tested surface revealed changes on the polished surface (Fig. 3). It was damaged and several adhered alumina particles of the micrometer size were visible. Consequently, only the apparent mass losses were measured.

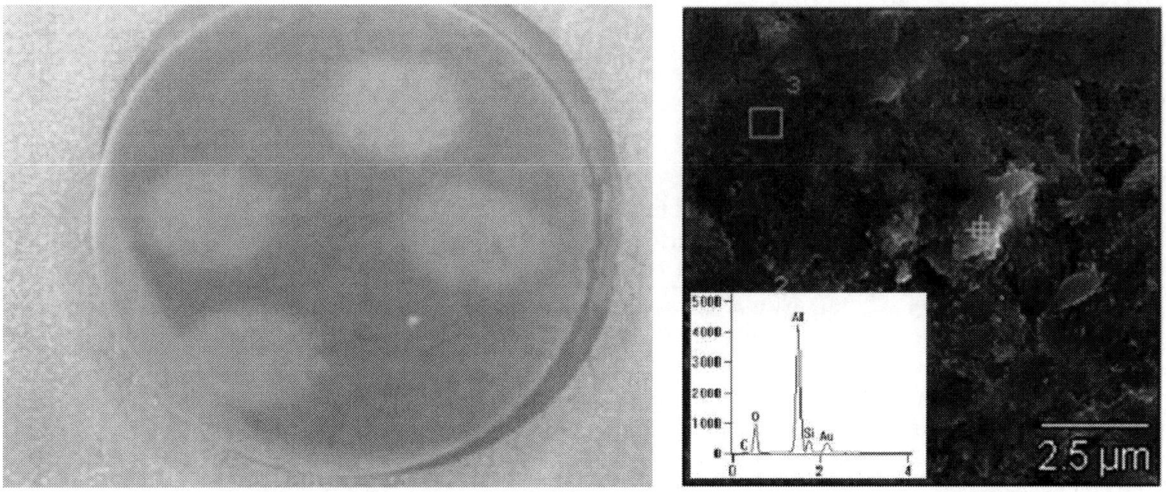

Figure 3. Surface of sample sintered at 1700/1500 °C for 360 minutes after erosion test; a) optical microscope, SEM image with EDS results in point 1 (insert)

The results of the wear resistance by the pin on disk after 480 m sliding distance showed very high wear rate in the range of 8.26-8.30 $*10^{-7}$ [mm^3 N^{-1} m^{-1}]. Figure 4 shows the 3D surface image of the specimen sintered at 1500 °C for 240 minutes after the wear test. In spite the very subtle differences in wear resistance we could conclude that elongation of the sintering time decreased the wear rate.

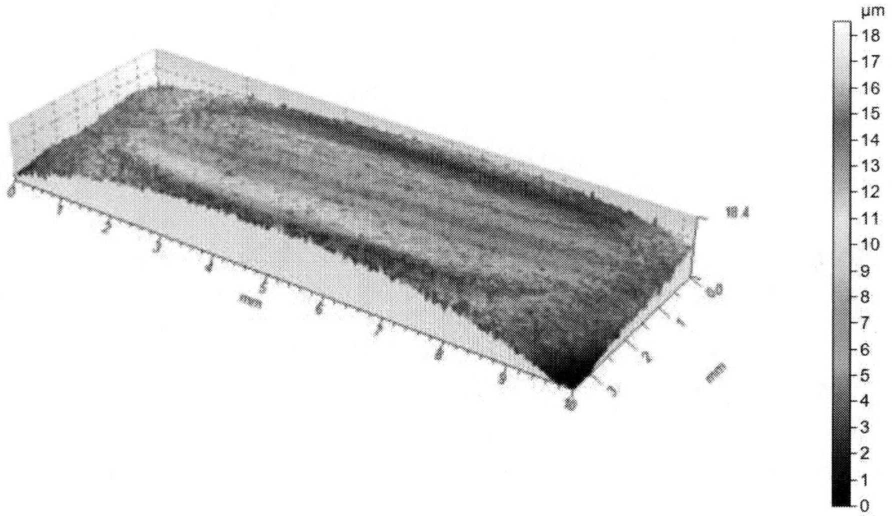

Figure 4. 3D image surface of sample after Pin on Block abrasion resistance test.

Wear results illustrate that despite the large number of cycles and various load, material's wear is very low. Roughness parameters of material's surface indicate a fine grain structure. Friction coefficient for tested samples was in the range of 0.08-0.12. The greatest stability of this coefficient was observed for the specimen sintered in the two-step mode (1700/1500 °C). Comparison of the obtained wear rate results with those reported in literature is risky as the wear process is closely related to the microstructure features and wear process itself. Nevertheless, obtained results fell into the upper range of wear resistance reported lately for nano or fine-grained silicon nitride ceramics manufactured by SPS technique [1-2, 19-20].

4. Summary

It has been shown, that application of MCP with moderate energy allowed pressureless sintering of silicon nitride ceramics and led to fine-grained microstructure and high density. The resultant material is characterized by dense, fine grained, homogenous structure. Further in terms of density no differences were found for the application of isothermal densification at 1500 °C or the application of two-step mode of densification. However, bending strength results showed that the two step densification mode (launching the high-temperature start of densification followed by the process at 1500 °C) produced less defective microstructure and better mechanical properties. This material is also characterized by high fracture toughness (K_{IC}), comparable with the commercial materials produced by the hot-pressing technique with microstructure of elongated β-Si_3N_4 grains.

Acknowledgement

The presented results are part of the project funded by Polish Ministry of Science and Higher Education No. N N507 373335 “Nanostructured silicon nitride based ceramics”.

References

[1] Gonzales-Julian J, Schneider J, Miranzo P, Osendi M.I and Belmonte M 2011 *J. Am. Ceram. Soc.* **94** 2542.
[2] Tatarko P, Kasiarova M, Dusza J, Morgiel J, Sajgalik P and Hvizdos P 2010 *Wear* **269** 867.
[3] Niihara N 1991 *J. Ceram. Soc. Jpn.* **99** 974.
[4] Gleiter H 2000 *Acta Mater.* **48** I-29.
[5] Herrmann M, Schultz I and Zalite I 2004 *J. Eur. Ceram. Soc.* **24** 3327.
[6] Bulić F I, Zalite I and Zhilinska N 2004 *J. Eur. Ceram. Soc.* **24** 3303.
[7] Nishimura T, Mitomo M, Hirotsuru H and Kawahara M 1995 *J. Mater. Sci. Lett.* **14** 1046.
[8] Shen Z and Nygren M 2011 J. Mater. Chem. 11 201.
[9] Kim H-D, Park Y-J, Han B D, Park M W, Bae W-T, Kim Y-W, Lin H T and Becher P F 2006 *Scr. Mater.* 54 615.
[10] Xu, X, Nishimura T, Hirosaki N, Xie R-J, Zhu Y, Yamamoto Y, Tanaka H 2005 *J. Am. Ceram. Soc.* 88(4) 934.
[11] Sopicka-Lizer M, Tańcula M, Włodek T, Rodak K, Huller M, Kochnev V, Fokina E and MacKenzie K 2008 *J. Eur. Ceram. Soc.* **28** 279.
[12] Xin Xu, Renli Fu and Ferreira J M F 2004 *Ceramics Int.* **30** 745.
[13] Xu X, Oliveira M I L L, Fu. Renli and Ferreira J M F 2003 *J. Eur. Ceram. Soc.* **23** 1525.
[14] Pawlik T, Sopicka-Lizer M, Michalik D, Włodek T 2011 *Arch. Metall. Mater.* **56** 1205.
[15] Anstis G R, Chantikul P, Lawn B R and Marchal D B 1981 *J. Am. Ceram. Soc.* 64 533.
[16] Czichos H, Klaffke D, Stantner E, Woydt M 1995 *Wear* **190** 155.
[17] Riley F L 2000 *J. Amer. Ceram. Soc.* **83** 245.
[18] Petzow G and Herrmann M 2002 *Struct. Bond.* **102** 47.
[19] Kurama S, Schultz J and Herrmann M 2011 J. Eur. Ceram. Soc. 31 921.
[20] Kim J-H, Kumar B V M, Hong S-H and Kim H-D 2010 *J. Am. Ceram. Soc.* **93** 1461.

Aspects of fabrication aluminium matrix heterophase composites by suspension method

A J Dolata, M Dyzia
Department of Materials Technology, Silesian University of Technology, Katowice, Poland

E-mail: anna.dolata@polsl.pl and maciej.dyzia@polsl.pl

Abstract. Composites with an aluminium alloy matrix (AlMMC) exhibit several advantageous properties such as good strength, stiffness, low density, resistance and dimensional stability to elevated temperatures, good thermal expansion coefficient and particularly high resistance to friction wear. Therefore such composites are more and more used in modern engineering constructions. Composites reinforced with hard ceramic particles (Al_2O_3, SiC) are gradually being implemented into production in automotive or aircraft industries. Another application of AlMMC is in the electronics industry, where the dimensional stability and capacity to absorb and remove heat is used in radiators. However the main problems are still: a reduction of production costs, developing methods of composite material tests and final product quality assessment, standardisation, development of recycling and mechanical processing methods. AlMMC production technologies, based on liquid-phase methods, and the shaping of products by casting methods, belong to the cheapest production methods. Application of a suspension method for the production of composites with heterophase reinforcement may turn out to be a new material and technological solution. The article presents the material and technological aspects of the transfer procedures for the production of composite suspensions from laboratory scale to a semi-industrial scale.

1. Introduction

Among structural materials the composites with an aluminium alloy matrix (AlMMC) are a group of materials which due to their properties (high specific elasticity modulus, high stiffness) are more and more frequently used in modern engineering constructions. Aluminium composites reinforced with hard ceramic particles (Al_2O_3, SiC) are gradually being implemented into production in automotive or aircraft industries, first of all due to high resistance to friction wear [1-4]. Another application of aluminium matrix composites is in the electronics industry, where the dimensional stability and capacity to absorb and remove heat is used in radiators [5]. Therefore, numerous research institutes conduct research to create efficient and cheaper methods of production of metal composites. Between the variety of processing techniques available for particulate reinforced metal matrix composites, stir casting is one of the methods accepted for the production and commercially practised [6,7,8]. The main problems are still: a reduction of production costs, developing methods of composite material tests and final product quality assessment, standardization, development of recycling and mechanical working [9].

2. Conditions of producing of composite suspension

Is a lot of works concerning develop different structural and technological solutions to introducing of particle reinforcement in the melt by the using the mixing method [4-8]. It was found that the amount of particles introduced into the liquid metal depends on their surface and the wetting angle [10]. The effectiveness of mechanical stirring, which has an effect on distribute the ceramic particles throughout the melt, depends on the components material, the stirrer geometry and its position, the speed of stirring, and also the composite mixture temperature [7]. That is why, from the standpoint of technology production of composite suspension by the vortex method should take into account many important factors such as:

- the matrix alloy's preparation (refining, modification of chemical composition, overheating),
- the ceramic particles preparation (chemical preparation or thermal treatment)
- the intensity of stirring the composite suspension,
- the homogenization conditions (time, temperature)- important for uniformly distribution of reinforcement in the liquid metal matrix.

For many years, the research relating to the manufacture of composite suspensions are also in the area of authors study [11-13]. The results of such research carried out under the research and development project PBR N RO7 001106 the transfer production procedure of composite suspensions, developed at a laboratory scale by the authors, to a semi-industrial scale of production, so as to enable die-casting of air-compressor pistons in industrial conditions. The application of a suspension method for the production of composites with heterophase reinforcement may turn out to be a new material and technological solution [14-17].

3. Materials for research

The ceramic powders surface reinforcement were used to produce heterogeneous suspensions of silicon carbide and glassy carbon powders are shown on Figures 1 and 2. The typical microstructure of the applied aluminium matrix alloy are presented in Figure 3. The application of SiC and glassy carbon particles enables stabilizing the friction coefficient and, most of all, reducing the wear of the friction partner [15]. These properties may be used in a tribological piston-cylinder system, from which dimensional stability, strength and durability are required [12,13,17].

Figure 1. Surface of SiC ceramic powders used as reinforcement.

Figure 2. Surface of glassy carbon powders used as reinforcement : a) lamellar particles and b) spherical particles.

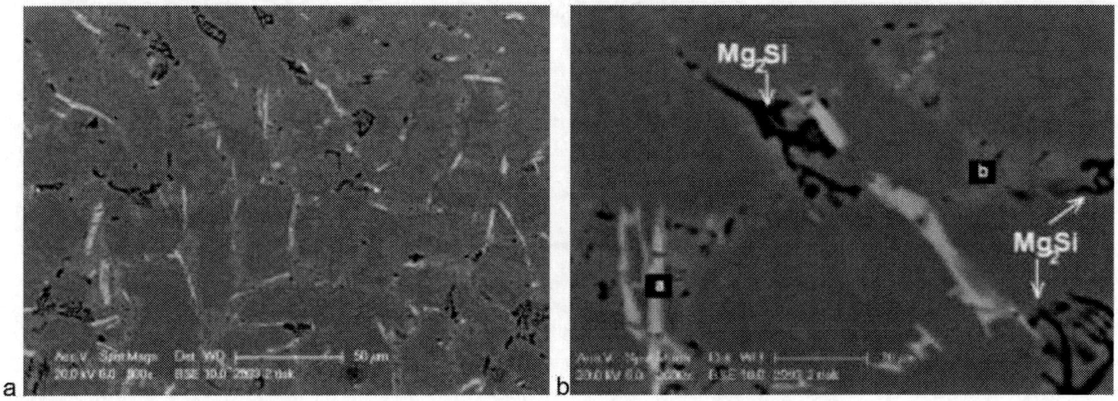

Figure 3. Structure of AlSi7Mg aluminium matrix alloy.

4. Experimental researches

The aim of research carried out under the research and development project PBR N RO7 001106 the transfer production procedure of composite suspensions, developed at a laboratory scale by the authors, to a semi-industrial scale of production, so as to enable die-casting of air-compressor pistons in industrial conditions. So far the process was conducted at two stages. Before the introduction of ceramic reinforcement into the matrix, the alloy composition was modified by adding Mg and Sr. Ceramic particles (mixture of SiC and glassy carbon) were preheated at a temperature of 350 °C and next, introduced into the liquid metal at 720 °C. At the second stage, the crucible with the composite suspension was placed in a hermetic chamber which facilitates degassing and homogenization under reduced pressure conditions. This way, 2.5 kg of ingot was obtained, which then was remelted and stirred for 30 minutes in the production conditions of the foundry. At this stage of tests, series consisting of five composite pistons were cast (fig. 4) in metal mould [16]. That technological tests confirmed the possibility of formation composite pistons from utilization the technology of mould casting. Research works carried out at the Department of Materials Technology, Silesian University of Technology have also confirmed good casting properties of heterophase composite suspensions [18]. In the study, tests of evaluate the ability the composite suspension to flow in the canal mold cavity were performed. In the research the spiral fluidity test, recording the temperature of the molten metal flowing through the system of caster and channel form in 1 and 5 of section were used (measurement channel with a length of 20 cm – fig. 5). Tests carried out at different temperatures of casting into mould (in the temperature range 720 ^{0}C to 680 ^{0}C). The estimated in this way the velocity of suspension flow at a temperature of 720 ^{0}C cast was approximately 42 cm s^{-1}, which is sufficient to fill the mold.

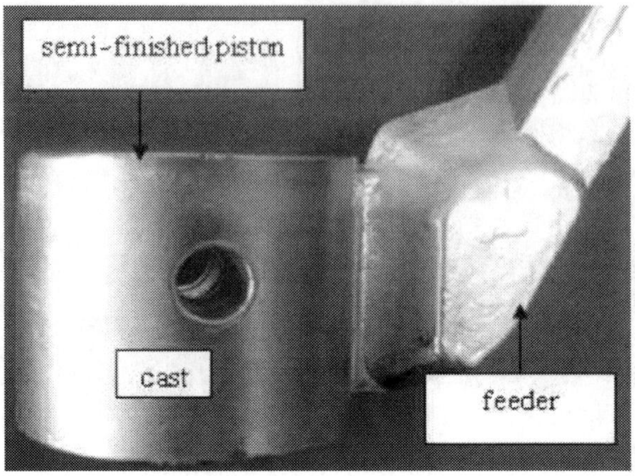

Figure 4. Semi-finished composite piston casted in metal mould.

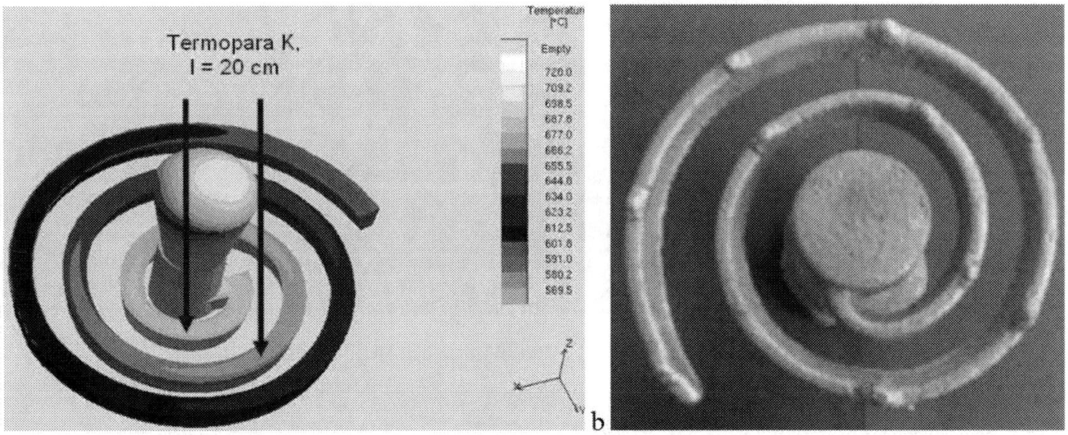

a b

Figure 5.. Schematic view of the measuring system and cast composite in the spiral fluidity test.

New equipment designed at Silesian University of Technology make possible to obtain approximately 50 kg of composite suspension in one cycle, in the joint processes of metal preparation, introduction of reinforcement and homogenization (fig. 6). The stand consists from hermetic chamber of furnace in which the crucible is built, as well as the movable cover with mixing system and feeding of ceramic particle unit. Composite suspension is obtained in several stages. After melting of the alloy the process of refinement is put on, with applying of the argon blow-by liquid metal from utilization the graphite stirrer. Next the additions of master alloy are introduced then the chemical constitution of the matrix alloy, and in next stage the ceramic particles are introduced. By the control of rotational speed of the feeding screw the velocity of introducing the reinforcement particles on liquid metal is limited. In the last stage the process of homogenizing suspension and the degassing in conditions of reduced pressure is put into [17]. In the first phase of testing a composite suspension was made using parameters established in previous studies carried out on a laboratory scale. As the base of matrix alloy was applied AlSi7Mg alloy. After melting and preheated at temperature 720 ^{0}C alloy has been subjected to the fifteen minutes process of refinement (the flow of argon 0,5 dm^3 h^{-1}). Then the master alloys AlMg25 and AlSr were introduced to liquid metal enlarging the part of magnesium in melt about 1% weights and modifying chemical composition with addition of 0.03 % Sr. After hold at temperature

350 °C SiC particles were poured into hopper. Particles were introduced in vortex area of liquid metal with velocity approx. 4 kg h^{-1}.

Figure 6.. Autoclave furnace for composites suspension production.

Homogenization of suspension was subjected in reduced pressure ($500 \cdot 10^2$ Pa) in 60 minutes time. All stages of preparation the suspension were performed in a protective atmosphere of argon. After this process, the furnace chamber was opened and began the process of hand casting of pistons with a diameter of 65 mm with applying a five-core metal mould. During the casting suspension was mixed with a pouring cup. Also in that stage the protective gas atmosphere was provided. Under the same conditions was prepared the heterophase composites reinforced with SiC and glassy carbon (Cg) particles (fig. 7).

Figure 7. Semi-finished composite piston cast to permanent mould machine GM110.

Amount of obtained composites suspension allows the use permanent mould casting machine GM110 working in the casting line "Złotecki" company. All technological parameters of casting process were according to productive standards non reinforced pistons. Carried out tests confirmed the possibility of forming composite pistons with applying casting technology. The view of semi-finished composite piston cast to permanent mould machine and its microstructure are presented on Figures 7 and 8.

IOP Conf. Series: Materials Science and Engineering **35** (2012) 012020 doi:10.1088/1757-899X/35/1/012020

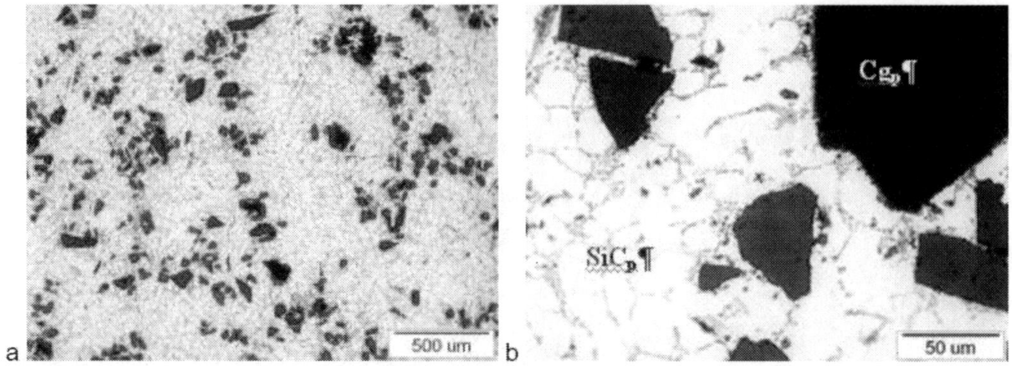

Figure 8.. Reprezentative microstructure of heterophase composite piston.

First of all, confirmed the major assumptions used in the construction of the furnace stand. Also confirmed the stability of the suspension and the possibility of casting composite pistons in trial batch series. In the manufacture of composite suspensions is very important to obtain uniform and stable distribution of reinforcement in the matrix. The structure analysis confirmed a uniform distribution of reinforcing particles in the matrix and a slight porosity in the cast (fig. 8).Number of obtained semi-finished composite pistons is sufficient to the next stage of the research, which are attempting to machining of piston surface. Hard ceramic SiC particles improve the tribological properties of the composite, but their presence complicates the machining of the cast. Application of glassy carbon should allows to reduce these negative effects. Preliminary tests at recessing the ring grooving confirmed the possibility of shaping the working surface of piston (fig. 9). Obtaining the proper shaping of piston require to select appropriate cutting tools and optimization of process parameters.

Figure 9.. AlSi7Mg/SiC+C heterophase composite piston after preliminary machining.

5. Summary

The repeatability of the shape of the materials at permanent mould machine casting significantly affect the ability of machining in CNC centers. Obtained composite suspension with particles of SiC reinforcement and a mixture of SiC and glassy carbon were characterized by good casting properties. Carried out technological tests confirmed the possibility of forming semi-finished composite pistons using permanent mould technology. Casting made in the industrial conditions characterized by a high repeatability. The produced a series of trial batch is sufficient to machining on automatic lathes. Currently, are carried out studies on the selection of appropriate cutting tools and machining conditions.

Acknowledgements
Scientific work carried out under the funding of the National Centre for Research and Development as part of the 6th competition of development projects , Project No. N RO7 001106.

References

[1] Koczak and all. 1993 Metal-Matrix Composites for Ground Vehicle, Aerospace, and Industrial Applications in Fundamentals of Metal-Matrix Composites, Edited by: Suresh, Subra; Mortensen, Andreas; Elsevier.
[2] Chawla N and Chawla K K 2006 *JOM* 67.
[3] Tung S C and McMillan M L 2004 *Tribol. Int.* **37** 517.
[4] Previtali B, Pocci D and Taccardo C 2008 *Composites: Part A* **39** 1606.
[5] Lemieux S, Elomari S, Nemes J A, Skibo M D 1998 J. Mater. Sci. 33 4381.
[6] Huang Y D, Hort N and Kainer K U 2004 *Composites: Part A* **35** 249.
[7] Balasivanandha Prabu S, Karunamoorthy L, Kathiresan S and Mohan B 2006 *J. Mater. Proces. Technol.* **171** Issue 2 268.
[8] Hashim J, Looney L and Hashmi M S J 1999 *J. Mater. Process. Technol.* **92/93** 1.
[9] Wieczorek J**,** Dyzia M. and Dolata A J 2012 *Solid State Phenomena* (in press).
[10] Sobczak N 2003 *Kompozyty (Composites)* **3** No 7 301 [in Polish].
[11] Dolata-Grosz A, Wieczorek J, Śleziona J and Dyzia M 2006 Archiv. Foundry 6 No 18 285.
[12] Dyzia M, Dolata-Grosz A, Śleziona J, Wieczorek J and Starczewski M 2007 *Composites* **7** No. 2 83.
[13] Dolata-Grosz A, Dyzia M, Śleziona J and Wieczorek J 2007 *Arch. Foundry Eng.* **7/1** 37.
[14] Dolata-Grosz A, Dyzia M and Lech-Grega M 2010 *Rudy i Metale Nieżelazne* **R55** No. 9 614 [in Polish].
[15] Wieczorek J, Śleziona J, Dolata-Grosz A and Myalski J 2008 *Kompozyty/Composites* **8** No.1 5.
[16] Dyzia M, Dolata-Grosz A, Wieczorek J and Śleziona J 2010 Arch. Foundry Eng. 10/1 301.
[17] Dyzia M 2011 *Solid State Phenomena* **176** 49.
[18] Dolata-Grosz A, Dyzia M and Śleziona J 2007 *J. Achiev. Mater. Manufact. Eng.* **20** Issues 1-2 103.

IOP Conf. Series: Materials Science and Engineering **35** (2012) 012021 doi:10.1088/1757-899X/35/1/012021

Impact of heat treatment on microstructure of steel 30X padded with wire G18 8Mn

Mirosław Łomozik*, Janusz Adamiec[1]**
* Institute Of Welding, Bl. Czeslawa 16-18, 44-100 Gliwice, Poland
**Department of Materials Science,, Silesian University of Technology University, Krasińskiego 8, 40-019 Katowice, Poland

E-mail: Janusz.adamiec@polsl.pl

Abstract. The practical purpose of padding is to increase mechanical properties, hardness and abrasion resistance of surface layers of structural elements while maintaining good plastic properties of the core, apply corrosion-resistant layers, regenerate and/or repair the surface etc. The article describes the results of macroscopic and microscopic metallographic examination of components made of 30X steel, which is an equivalent of 32HA steel produced in Poland, and surfaced with G18 8Mn filler wire using MAG method. One of components was not heat treated after surfacing while the other after surfacing was subjected to quenching and tempering. The examination was aimed at determining the effect of heat treatment on changes in the structure of surfaced components. The metallographic examination has been conducted using optical microscopy in the area of parent metal of 30X steel as well as in heat affected zone (HAZ) in steel under and in the surfacing weld. HAZ width under the surfacing weld has also been measured. On the basis of the results showed that heat treatment after padding is responsible for the presence of the structure of tempered martensite in the heat affected zone and is responsible for a greater width of the HAZ, if compared with the condition without heat treatment.

1. Introduction

According to a definition found in related reference publications, padding is applying a layer of molten metal on a metal object by means of a welding method, with the penetration of the base surface [1]. The practical purpose of padding is to increase mechanical properties, hardness and abrasion resistance of surface layers of structural elements while maintaining good plastic properties of the core, apply corrosion-resistant layers, regenerate and/or repair the surface etc. The higher the content of carbon (above 0.25%) and alloying elements, such as chromium and molybdenum, in the steel being padded, the more problematic the padding process and the higher the hardening tendency in the heat affected zone (HAZ); the latter trend being attributable to the hardening of the HAZ as a result of a martensitic transformation.

In many cases, elements after padding are subjected to heat treatment, the purpose of which is to obtain the properties of a padding weld and base metal as required by operational conditions.
This article was inspired by the necessity to determine the impact of heat treatment on the microstructure of steel subjected to padding.

---

[1] To whom any correspondence should be addressed.

Published under licence by IOP Publishing Ltd

IOP Conf. Series: Materials Science and Engineering **35** (2012) 012021 doi:10.1088/1757-899X/35/1/012021

2. Subject and purpose of research

Research-related tests involved the production of padding welds on pipes of an external diameter of 180 mm and weld thickness of 26 mm, made of steel X30 acc. to the standard GOST 4543-71, the domestic equivalent of which is steel grade 32HA acc. to the standard PN-89/H-84023/07/A1 [2]. The chemical composition of the steels 30X and 32HA is presented in Table 1. Prior to padding, the pipes underwent hardening and tempering.

Table 1. Chemical composition of fragments of steel 30X (32HA).

Steel	Content of chemical element in %								
	C	Mn	Si	P	S	Cr	Ni	Cu	Al
30X	0.30	0.68	0.27	0.013	0.005	0.96	0.08	0.09	0.029
32HA	0.28 - 0.35	0.55 - 0.85	0.17 - 0.37	max. 0.025	max. 0.025	0.85 - 1.15	max. 0.20	max. 0.20	min. 0.02

A single-layer padding weld was made by means of the MAG method (135), applying the following ranges of electrical parameters of the process: welding current I = 180-200 A, electric arc voltage U = 23-24 V. A mix designated as M13 (Ar + 2% O_2) was used as shielding gas - following the requirements of the standard PN-EN ISO 14175-M13-ArO-2 [3]. Padding was carried out with a wire G18 8Mn (diameter 1.2); following the requirements of the standard PN-EN ISO 14343-A-G 18 8 Mn [4]. The chemical composition of the weld deposit of the wire G18 8MN is presented in Table 2.
The process of padding was carried out in a manner that would allow obtaining the thickness of a padded layer of between 2 and 3 mm. The tests involved the use of padded pipes designated as PN and POC respectively (Fig. 1a). The area of test-related sampling is presented in Figure 1b.

Table 2. Chemical composition of weld deposit of wire G18 8Mn

Weld deposit	Content of chemical element in %							
G18 8Mn	C	Mn	Si	P	S	Cr	Ni	Cu
	0.067	7.03	0.78	0.019	0.005	19	8.94	0.10
	Mo	Nb	Ti	Co	V	Al	B	N
	0.15	0.01	0.01	0.04	0.06	0.005	0.0009	0.054

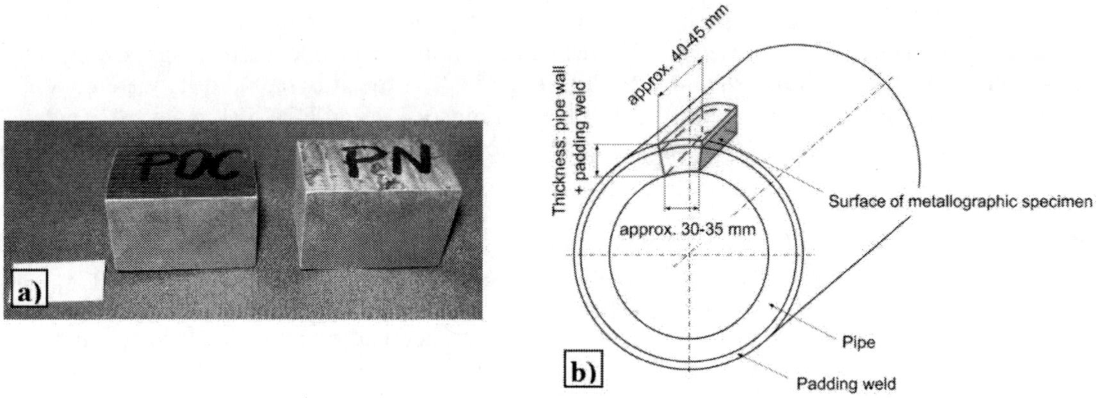

Figure. 1. Fragments of pipes with padding welds: a) main view, b) sampling manner and dimensions of elements being tested

The fragment sampled from the pipe not subjected to heat treatment after padding was designated as PN. In turn, the fragment designated as POS was sampled from the pipe which underwent hardening and tempering after padding (Fig. 1a). During hardening the temperature of the test piece was 890°C, the hold time - approx. 1.5 hours, cooling in water; during tempering the temperature of the test piece was 490°C, the hold time was approx. 3 hours, cooling in water. The purpose of the tests was to determine the impact of heat treatment (hardening and tempering) on the structure of the area subjected to padding.

3. Course and results of tests

Metallographic specimens for macro- and microscopic tests were prepared on the surfaces along the main axis of the pipe (Figure 1b). The surfaces of the metallographic specimens were prepared following the requirements of the standard PN-EN 1321 [5]. In order to reveal the macrostructure of the padding weld, the test pieces were etched in Adler etchant. Examples of macrostructures are presented in Figure 2.

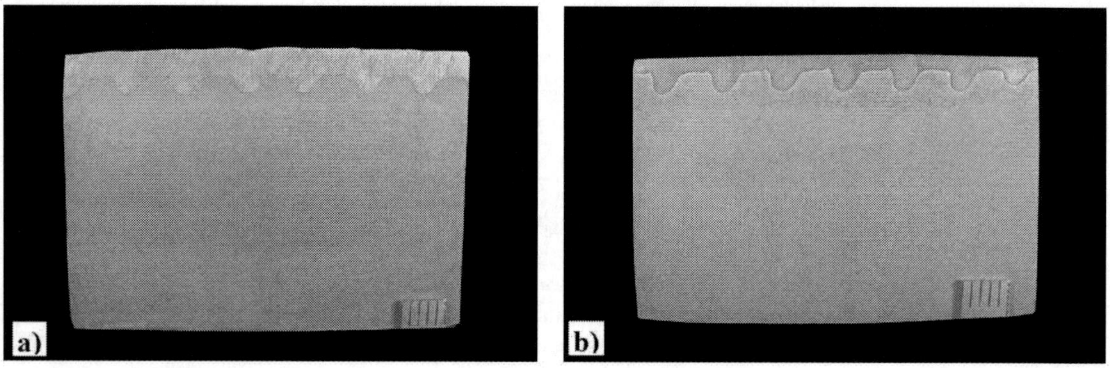

Figure. 2. Macrostructures of padding welds: a) test piece PN padded without heat treatment, b) test piece POC padded and heat treated

The quality of the padding welds in the pipe fragments being tested was evaluated following the requirements of the standard PN-EN ISO 15614-7 [6]. The test pieces did not reveal any crack or other unacceptable flat imperfections (defects).

Owing to significant differences in the content of chromium, nickel, manganese and carbon between the padding weld metal and the steel of the pipe, it was necessary to apply double-stage etching. At the first stage the microstructure of the steel was revealed by means of the etchant Nital and, afterwards, electrolytic etching was applied to reveal the microstructure of the padding weld. The electrolytic etching involved the use of the aqueous solution of nitric acid (60% HNO_3 + 40% H_2O), etching parameters: U = 4 V, t = 2 s. The microscopic metallographic tests were carried out by means of a Leica-manufactured light microscope MeF4M collaborating with a computer supporting Buehler-developed measurement software OMNINET Enterprise. The metallographic tests were conducted in the characteristic areas of the padding weld of the test pieces PN and POC i.e. in the base metal of the pipe steel, heat affected zone (HAZ) and in the padding weld. The results of the microscopic tests are presented in Figures 3-8.

The results of measurements of the width of the HAZ area, obtained by means of the measuring module of the programme OMNIMET Enterprise are presented in Table 3. The microstructural tests were supplemented by the microanalysis of chemical composition, carried out using a scanning electron microscope HITACHI S-4200, provided with an X-ray microanalysis system NORAN Voyager 3500 and an electron dispersion spectroscope. Chemical composition was tested at voltage accelerating an electron beam of 25 keV. The results of surface tests and linear distributions of chemical elements on the padding weld fusion line are presented in Figures 8-9.

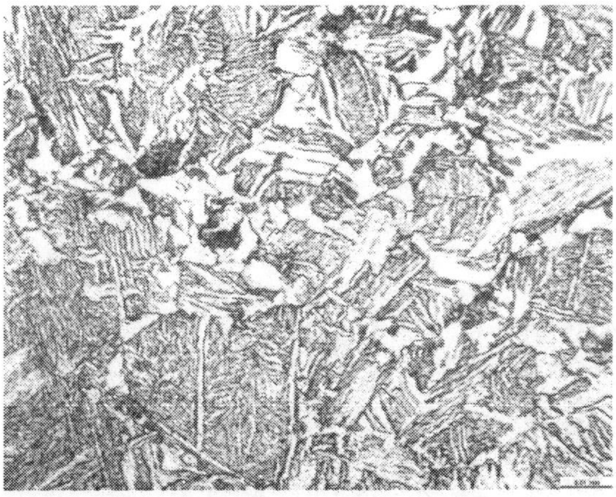

Figure 3. Base metal of steel 30X, bainitic structure with granular and lamellar ferrite and slight areas of pearlite, test piece PN, etch. Nital, mag. 1000×

Table 3. HAZ width values in test pieces PN and POC

Test piece	HAZ width, mm													
	1	2	3	4	5	6	7	8	9	10	11	12	13	**av.**
PN	1.905	1.960	3.142	2.700	2.823	1.513	1.914	3.040	2.879	2.109	1.626	2.600	1.854	**2.313**
POC	1.863	3.310	3.209	3.488	3.300	2.458	1.945	2.533	2.440	1.721	2.480	3.515	3.480	**2.749**

4. Analysis of results and conclusions

The microstructure of the steel 30X in the test pieces designated as PN and POC is a mixture of bainite, granular ferrite and/or lamellar ferrite located mainly on grain boundaries of former austenite and of slight amounts of pearlite (Fig. 3). In case of the test piece POC, heat treatment conducted after padding does not cause changes in the microstructure of the base metal of the steel 30X, as opposed to the test piece PN.

In the heat affected zone of the steel 30X in the test piece PN one can notice typical structural areas characteristic of steel with a higher carbon content. The said areas, located in the direction from the fusion line deep inside the base metal (Fig. 4a), are as follows:

- coarse-grained superheated area with martensite and slight amounts of ferrite (Fig. 4b),
- coarse-grained area heated in the temperature range A_{C1}-A_{C3}, containing a mixture of martensite, bainite and slight amounts of ferrite (Fig. 4c),
- fine-grained areas (normalisation and incomplete normalisation), containing a mixture of ferrite, pearlite and slight amounts of bainite (Fig. 4d, e).

The microanalysis of chemical composition, carried out in the padding weld fusion line revealed changes in the chemical composition (Fig. 8) i.e. an increase in the content of chromium, nickel and manganese. This phenomenon is connected with the mixing of the welding wire metal and molten base metal, and is confirmed by the linear distributions of chemical elements in the fusion line (Fig. 8 d, e).

The heat treatment, consisting in hardening and tempering, carried out after padding, caused changes in the HAZ areas of the steel 30X in the test piece POC (Fig. 6a). In the HAZ area, near the fusion line one can observe bainite and granular ferrite located mainly on the boundaries of former austenite - Fig. 6b. In the HAZ, after the heat treatment, similarly as in the test piece PN, it was possible to observe areas having the structure of coarse-grained tempered martensite - Fig. 6c.

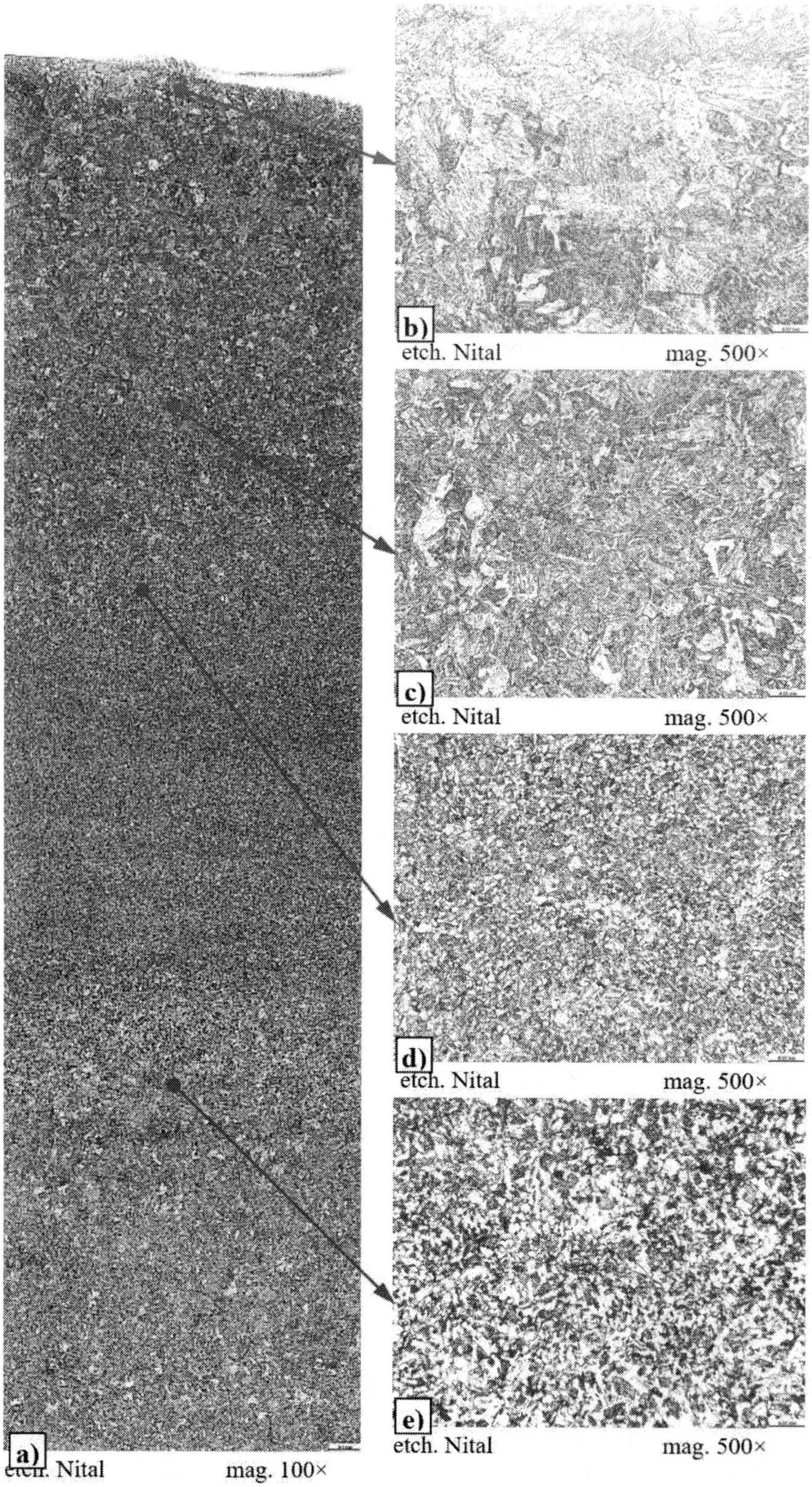

Area of coarse-grained structure near fusion line. Martensite + slight amounts of ferrite

Area of coarse-grained structure heated in temperature range A_{C1}-A_{C3}. Mixture of martensite, bainite and ferrite

Area of fine-grained structure (normalisation area). Mixture of ferrite, pearlite and slight amounts of bainite

Area of fine-grained structure (incomplete normalisation area). Mixture of o different-sized grain, ferrite, pearlite
and slight amounts of bainite

Figure 4. Main view of HAZ microstructure in the direction from fusion line (up) to base metal of steel 30X (down), test piece PN

IOP Conf. Series: Materials Science and Engineering **35** (2012) 012021 doi:10.1088/1757-899X/35/1/012021

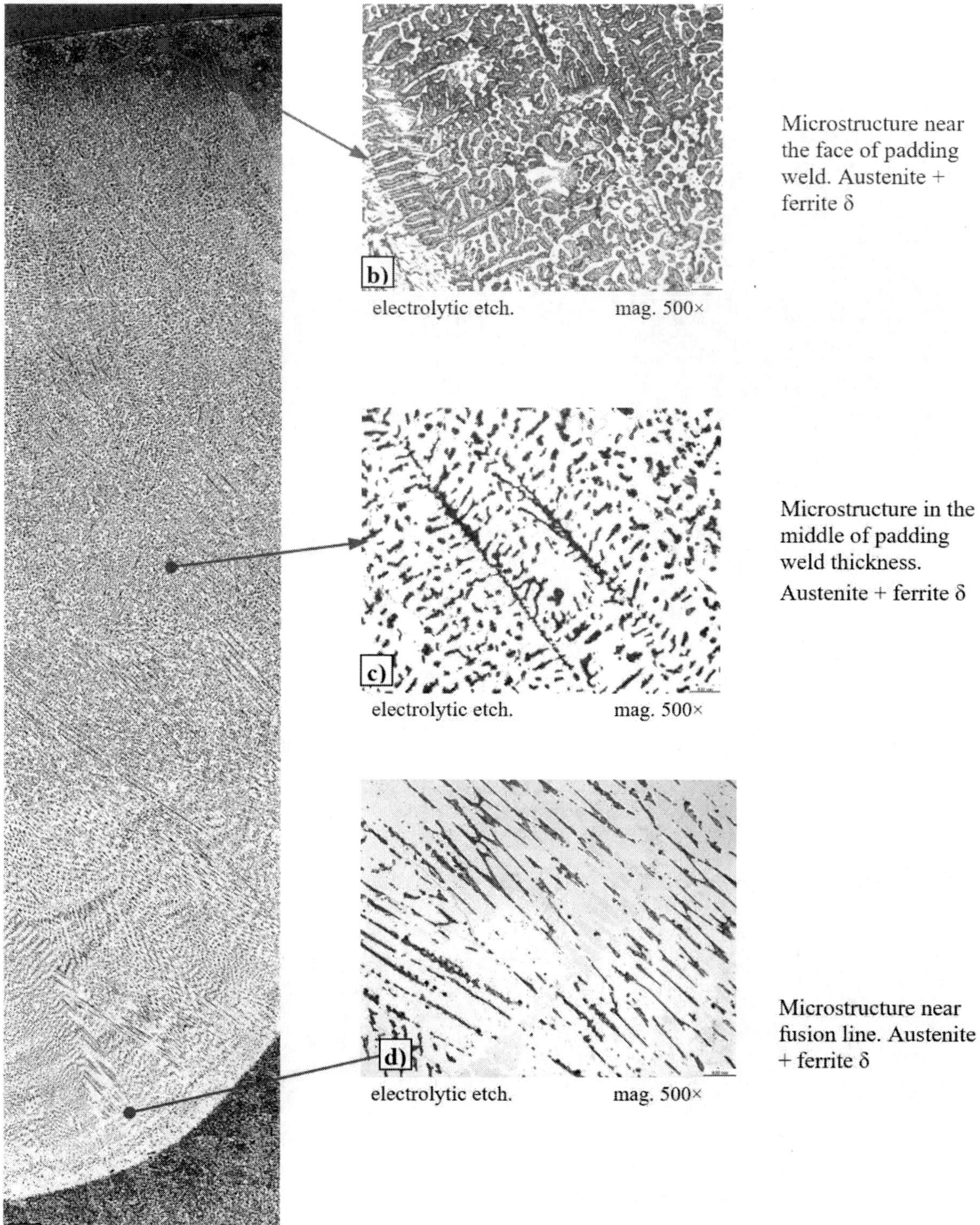

Figure 5. Structure of padding weld without heat treatment, test piece PN

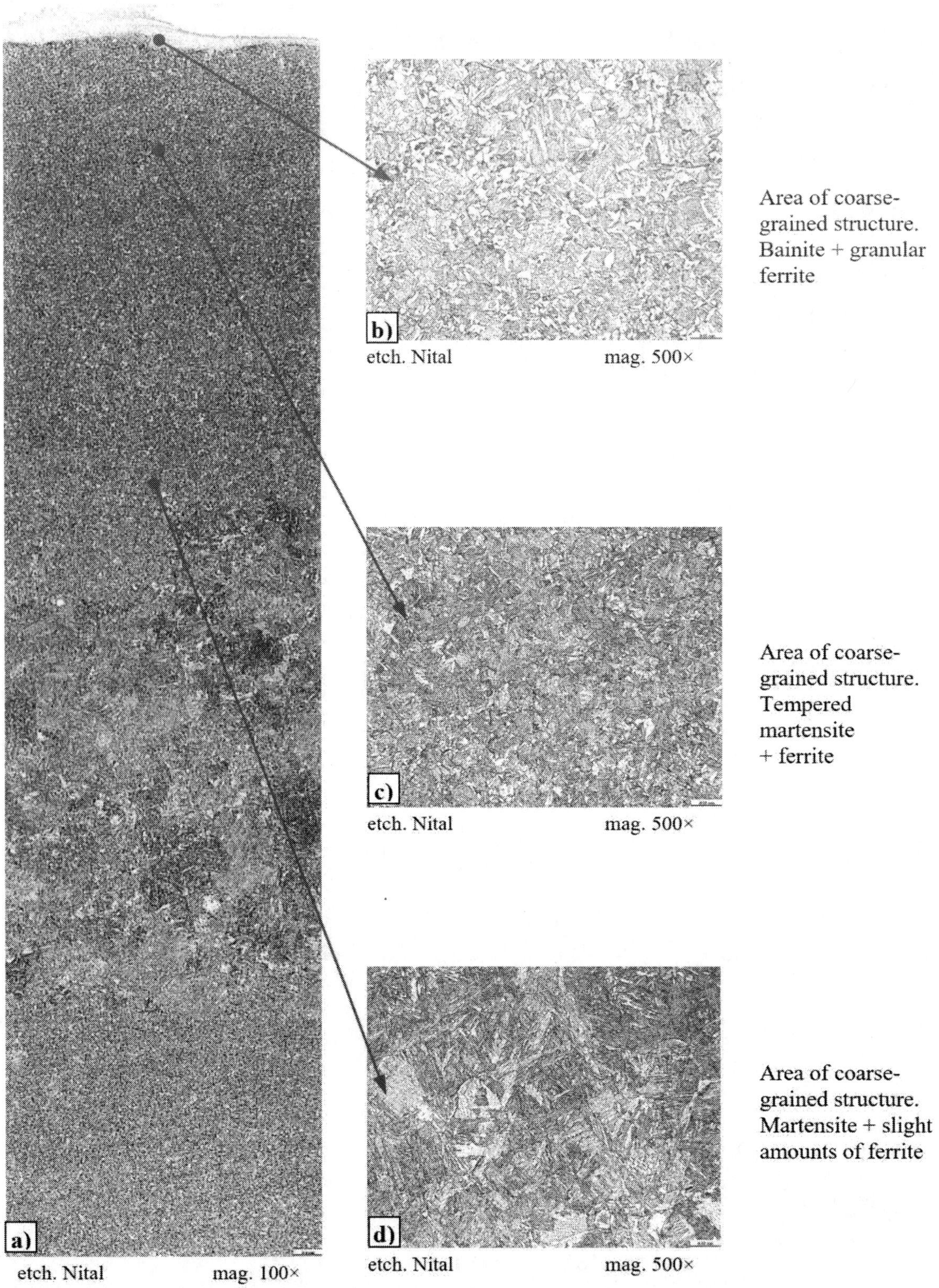

Figure 6. Heat affected zone in the direction from fusion line (up) to base metal of steel 30X (down), test piece POC

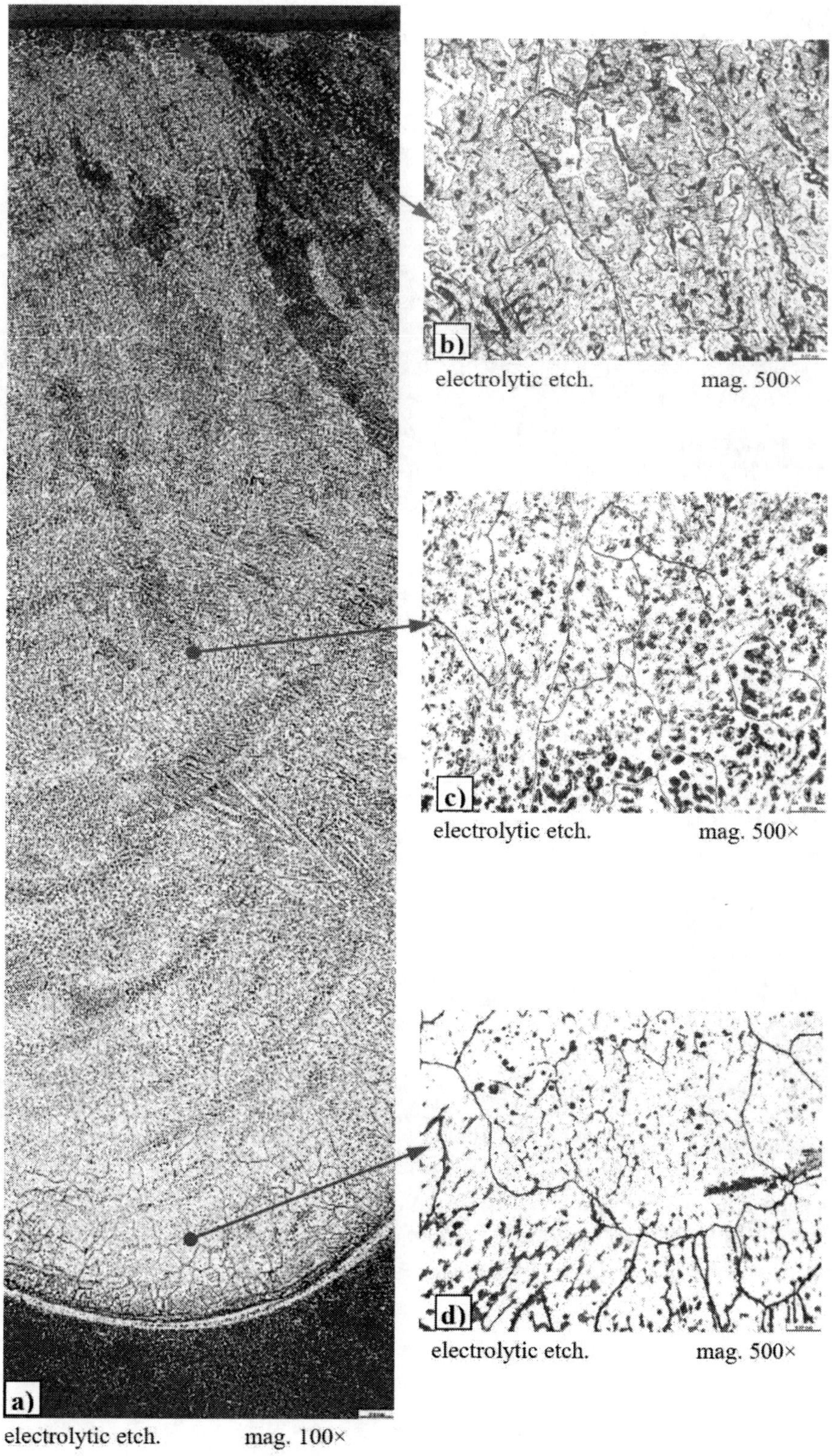

Microstructure near the face of padding weld. Ferrite δ + ferrite + austenite

Microstructure in the middle of padding weld thickness. Ferrite + ferrite δ

Microstructure near fusion line. Ferrite + ferrite δ

Figure 7. Structure of padding weld after heat treatment, test piece POC

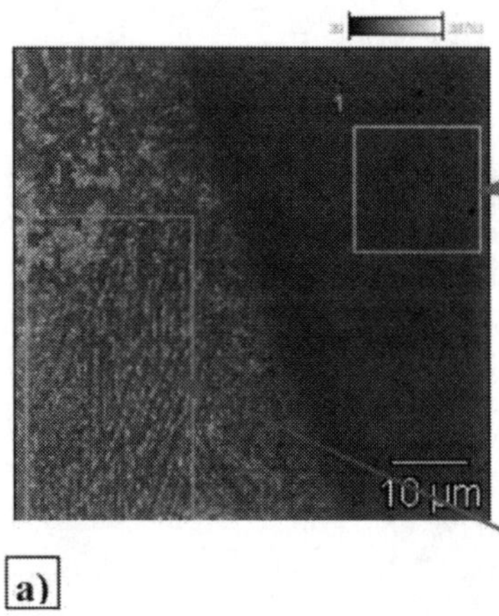

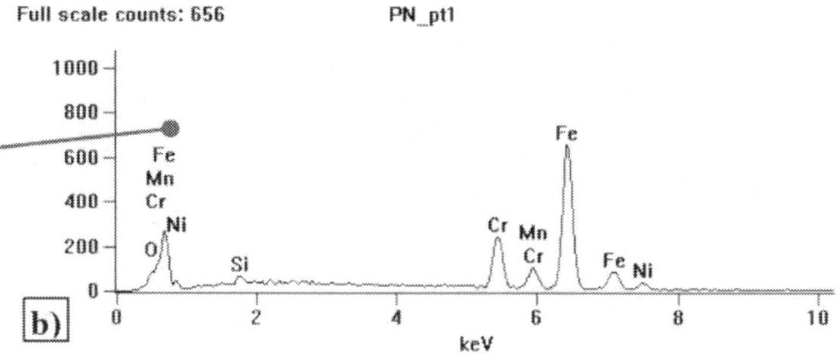

Padding weld

Weight %	*Si*	*Cr*	*Mn*	*Fe*	*Ni*
PN_pt1	0.74	14.33	6.27	73.40	5.26
PN_pt2	0.38	1.07	-	98.55	-
Atom %	***Si***	***Cr***	***Mn***	***Fe***	***Ni***
PN_pt1	1.45	15.14	6.27	72.22	4.93
PN_pt2	0.76	1.14	-	98.10	-

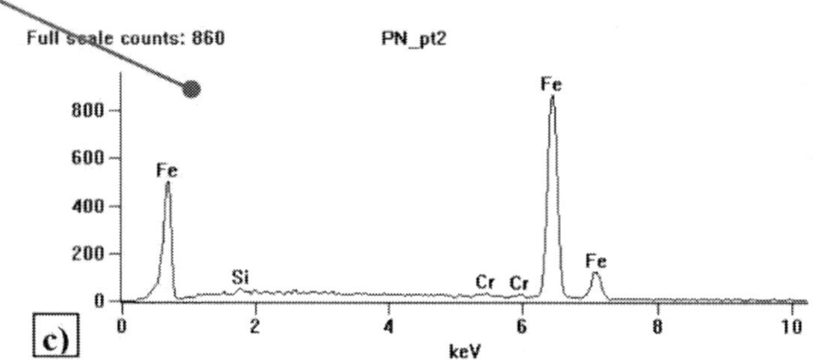

Heat affected zone

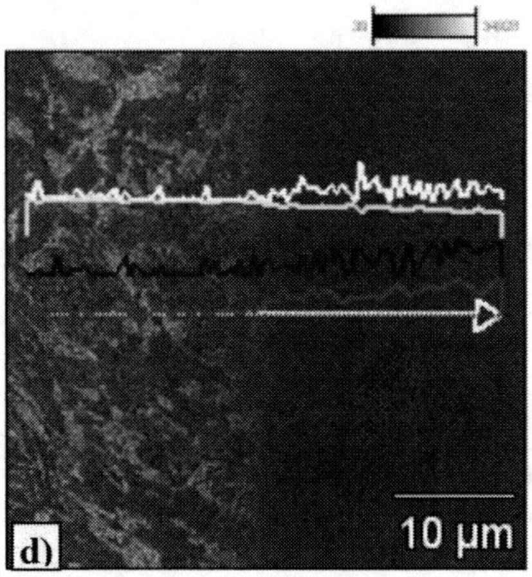

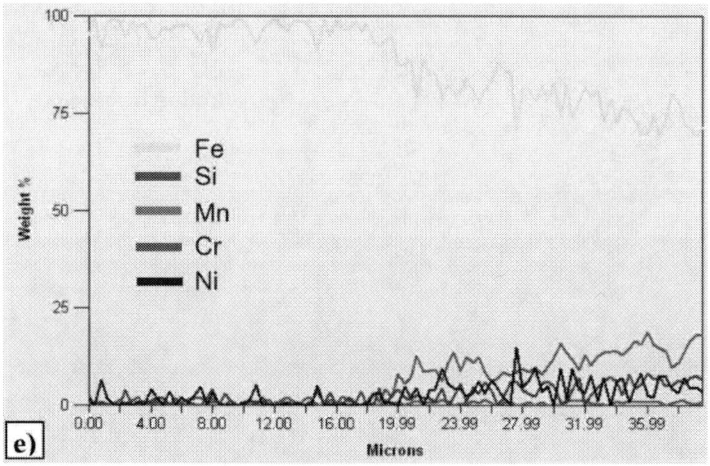

Figure 8. Results of microanalysis of chemical composition of padding area, test piece PN: a) test area, fusion line, b) EDS radiation spectrum - padding weld area, c) EDS radiation spectrum - HAZ area, d, e) linear distribution of chemical elements in fusion line

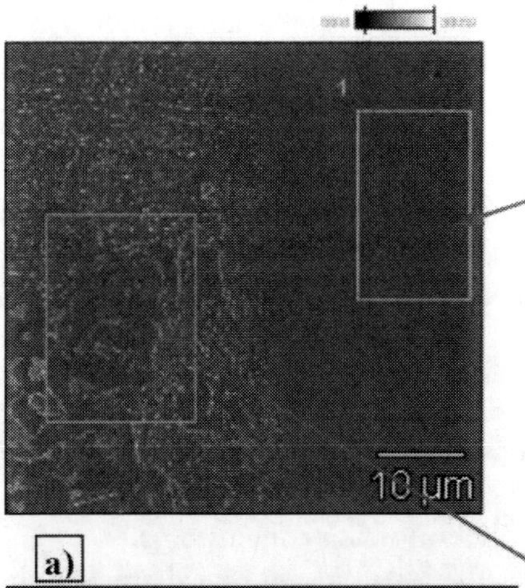

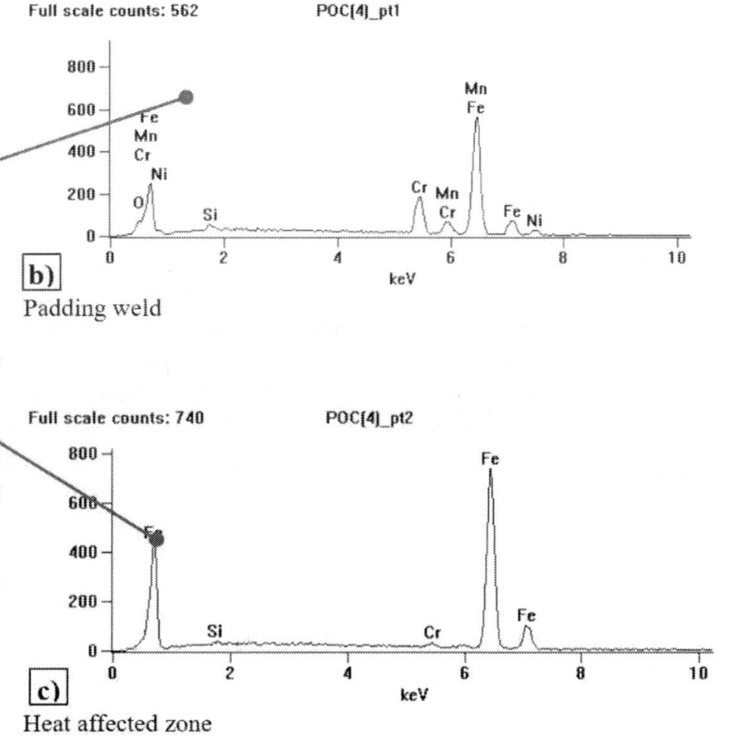

Weight %	*Si*	*Cr*	*Mn*	*Fe*	*Ni*
POC_pt1	0.58	13.31	5.19	74.92	5.99
POC_pt2	0.21	1.08	-	98.71	-
Atom %	***Si***	***Cr***	***Mn***	***Fe***	***Ni***
POC_pt1	1.15	14.10	5.21	73.92	5.62
POC_pt2	0.41	1.16	-	98.43	-

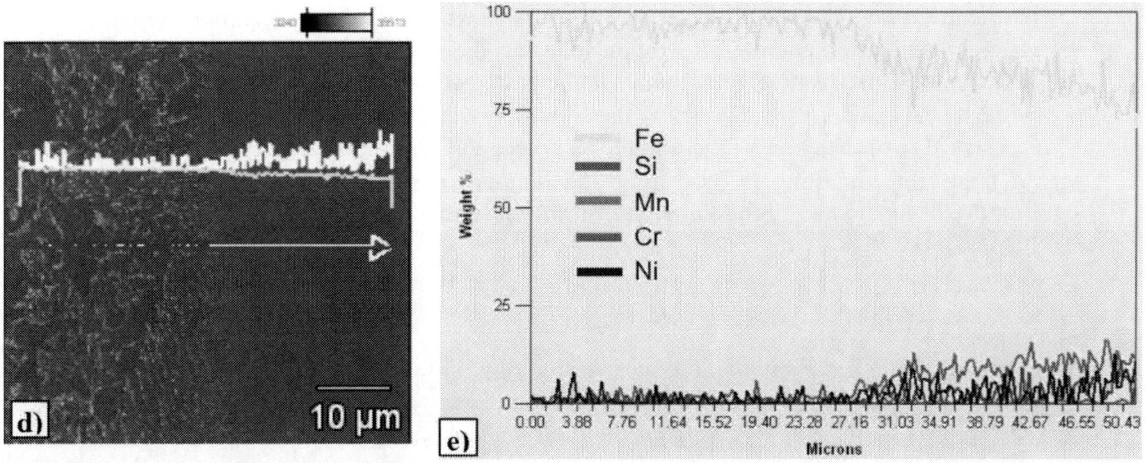

Figure 9. Results of microanalysis of chemical composition of padding area, test piece POC: a) test area, fusion line, b) EDS radiation spectrum - padding weld area, c) EDS radiation spectrum - HAZ area, d, e) linear distribution of chemical elements in fusion line

The results of the microanalysis of the chemical composition of the padding weld fusion area after heat treatment (POC) also reveal changes in the chemical composition in the fusion line (Fig. 9).The chemical composition of the padding weld area results from the mixing of the base metal of the steel 30X and welding wire G18 8Mn (Fig. 9 a-c). The linear distribution of chemical elements in the fusion line indicates a gradual transfer of chemical elements from the welding wire to the padding weld and a proper degree of their intermixing.

The measurements of the HAZ width in the test pieces before and after the heat treatment reveal that after hardening and tempering, the width of the HAZ area in the steel 30X slightly increases if compared to the state preceding the heat treatment (2.3 mm for the test piece PN and 2.7 mm for the test piece POC). (Table 3).
While analysing the chemical composition of the welding wire used for padding one can notice that it contains both austenite-generating elements such as manganese and nickel and a slight amount of cobalt as well as ferrite-generating elements such as chromium, silicon, molybdenum, niobium, titanium, aluminium and vanadium. Such a chemical composition of the filler metal is reflected in the structural composition of the padding weld in the test piece PN i.e. the one not subjected to heat treatment after padding. The microstructure of the padding weld of the test piece PN contains austenite and ferrite delta (Fig. 5). The heat treatment after padding causes changes in the padding weld morphology (Fig. 7). In the padding weld of the test piece POC one can observe the dominance of ferritic structure being the mixture of ferrite and high-temperature ferrite δ. Only the surface layer of the padding weld contains austenite (in addition to ferrite) (Fig. 7b).
On the basis of the tests described above, it was possible to draw the following conclusions:

- fragments of the pipes with the padding welds subjected to the tests did not reveal any welding imperfections which would disqualify or decrease the quality of the padding welds according to the standard PN-EN ISO 15614-7.
- structure of the base metal of the steel 30X (equivalent to the steel 32HA) in the test pieces PN and POC is composed of bainite, granular and/or lamellar ferrite located mainly on the boundaries of former austenite and of slight amounts of pearlite.
- heat treatment, consisting in hardening and tempering, does not cause any changes in the microstructure of the base metal of the steel 30X, if compared with the state present in the steel 30X in the test piece PN.
- heat affected zone in the test piece PN, after padding and without heat treatment is characterised by the presence of structural areas typical of steels which have higher carbon content.
- heat treatment after padding is responsible for the presence of the structure of tempered martensite in the heat affected zone of the test piece POC and is responsible for a greater width of the HAZ, if compared with the condition without heat treatment.
- microstructure of the padding weld in the test piece PN is composed of austenite and ferrite δ.
- heat treatment consisting in hardening and tempering is responsible for the presence of the mixture of ferrite and high-temperature ferrite δ in the microstructure of the padding weld of the test piece POC.

5. References

[1] Pilarczyk J *Engineer's guide. Welding technology. Vol. 2.* (Warszawa: WNT)
[2] PN-89/H-84023/07:1989/A1:1997 Steel for specific applications. Pipe steel. Grades
[3] PN-EN ISO 14175:2009 *Welding consumables. Gases and gas mixtures for fusion welding and allied processes*
[4] PN-EN ISO 14343:2010 *Welding consumables. Wire electrodes, strip electrodes, wires and rods for arc welding of stainless and heat resisting steels. Classification*
[5] PN-EN 1321:2000 *Destructive tests on welds in metallic materials. Macroscopic and microscopic examination of welds*
[6] PN-EN ISO 15614-7:2009 *Specification and qualification of welding procedures for metallic materials. Welding procedure test. Part 7: Overlay welding*

IOP Conf. Series: Materials Science and Engineering **35** (2012) 012022 doi:10.1088/1757-899X/35/1/012022

Ceramic or metallic? - material aspects of compact heat regenerator energy efficiency

M Wnek
Politechnika Śląska, Wydział Inżynierii Materiałowej i Metalurgii,
Katedra Metalurgii, ul. Krasińskiego 8, 40-019 Katowice

E-mail: mariusz.wnek@polsl.pl

Abstract. The metal industry cannot afford the financial mismanagement in the era of rising energy prices and thus, the high efficiency devices should be used. In the metallurgical thermal processes the combustion air temperature increasing is one of the methods for obtaining the heat transfer intensification and the furnaces efficiency rising. Therefore the new and effective heating technologies in thermal processes are demanded all the time. The regenerative systems are most effective in terms of the heated air level. The individual regenerators for burners are the newest solutions where the temperature of 1100 °C is reachable for the exhaust temperature of 1200 °C. Based on research results, performed for the assumed exhaust temperature of 1100 °C, the paper presents possibilities of changeable different materials using as a regenerator filling in the aspect of its operation efficiency. Such materials as high-temperature steel, Al_2O_3 and SiC have been considered. The paper presents the selected data research, dealing with the air combustion temperature obtained for the same type of regenerator filling of considered materials. The fuel consumption reduction and reduction of CO_2 emission, for metal regenerator filling, have been presented finally as an economic and environmental aspect accordingly to the air preheated.

1. Introduction

The high energy efficiency of thermal devices is required while increasing environmental restrictions in the era of high competition, rising energy prices and diminishing fossil fuel resources which are required to conduct of metallurgical technology. Therefore the world-metallurgical industry keeps looking for some new and effective solutions in thermal metallurgical processes which will be able to improve the heat transfer intensification and the furnaces efficiency. To achieve this, the most popular method is the raising of combustion air temperature. The air temperature increase causes the reduction of fuel consumption and decrease the fumes production and consequently the environment is better protected. However, the effectiveness of used method strongly depends also on the effective solutions as well. Thus, if very high combustion air temperature is needed, the regenerator system should be used. Generally, the high air preheating by means of a special kind of exchangers using is a new type of combustion and it is called the High Temperature Air Combustion (HTAC) or High Preheated Air Combustion (HPAC) [1-3]. A temperature of more than 1000 °C is obtainable in this system as indicated by some literature materials [1,4-7] which in turn significantly reduces the fuel consumption and CO_2 emissions [5,8].

The individual regenerators for burners are the newest solutions where the very high combustion air temperature is reachable [7,9]. However the specific design and appropriate materials are needed for using this technology and first of all, the system long life working should be ensured in the high temperature conditions. The HPAC technology is already applied in different kinds of furnaces by means of different fuel type such as gas, oil and even coal and biomass. This new technology provides the most effective heat recovery, reduction of energy consumption, cleaner and cheaper production.

The ceramic materials are the most commonly used as a filling for the modern heat regenerators. The innovative heat exchanger type which has become popular is the "honeycomb" [2,6,9]. This kind of regenerator is monolithic regular solid with ducts going along inside. A square shape of these ducts (in cross-section) is the most often used [1,2]. The new regenerative burner system with the "honeycomb" heat exchanger is called the High-cycle Regenerative Combustion System (HRS).

The regenerator filling type, as a honeycomb, is considered in this article. It is because of the growing popularity of that filling and documented, in the literature, significant effects of its use. However, due to the drastic structural parameters of filling, in some cases the wall thickness of around 0.4 mm is presented, it makes production of such filling costs very high. These costs primarily are a result of the use of high-quality manufacturing process and in that case the assistance in the form of funding is essential. Thus, it seems expedient to consider a different material - easier to process and cheaper, which could be a substitute for the ceramic material.

Thereby, in the case of heat technologies in which the exhaust gas temperature reaches a value up to 1150°C, a metal material can be considered as a regenerative filling. Due to the temperature level it is advisable to look into the application of heat-resistant steel such as nickel-chrominium steel.

The comparison of four ceramic materials and heat-resistant steel (as a substitute) for the construction of regenerator filling has been presented in this article. The selected ceramic type materials as Al_2O_3 and SiC have been considered. For comparison of the regenerator materials, the average heated air temperature, for the pseudo-steady state of regenerator, has been assumed. To demonstrate the quality of considered materials, the so-called quality indicator has also been introduced and then the possibilities and some benefits of interchangeable using of metal filling have been discussed as well.

2. Operation of regenerative burners

The burners usually work in pairs in regenerative systems as presented in Figure 1. The fluids are delivered and received by the fans – pumping and sucking. While burner B1 is firing, its exhaust is sucked through burner B2 into regenerator R2 where it emits heat to a storage medium. The exhaust temperature decreases during their flow through the regenerator, whereas the temperature of the regenerator increases. After a certain adjustable time period (it is called reversion time or blow period) the process will be reversed as in Figure 2. Burner B2 starts firing and its exhaust flows through burner B1 and regenerator R1. Cold air is heated up in regenerator R2 by the storage medium, which was heated during the previous cycle. The temperature of flowing air increases and temperature of regenerator R2 decreases. The air (preheated in that way) is delivered into burner B2 for combustion process. The whole process is repeated according to the reversion time which is usually taken in the range from 15 to 180 s.

It can be said that the regenerator is heated and cooled during the first and second half-period respectively, as described above, as well as the fluids flow direction is changed in each period. The regenerator works in an unsteady state. Its temperature as well as flowing fluids temperature are always a function of time and space. But for the users the most important is outlet temperature of fluids, particularly the outflow air temperature.

The average temperature can be specified for fluids in the outflow when the pseudo-steady state is achieved and then this temperature should be used for comparisons of the different heating type solutions or materials used for the regenerator filling.

IOP Conf. Series: Materials Science and Engineering **35** (2012) 012022 doi:10.1088/1757-899X/35/1/012022

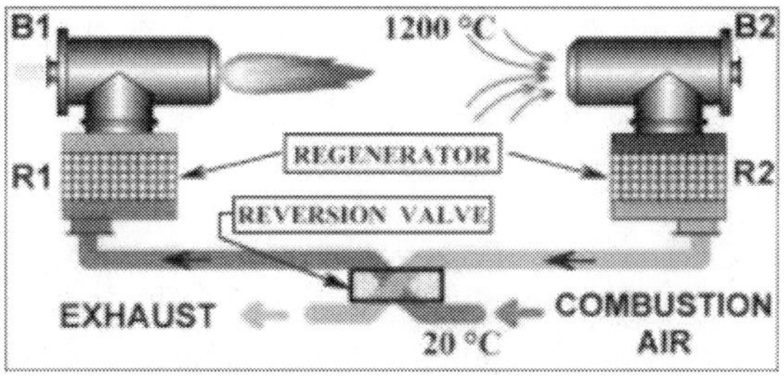

Figure 1. Regenerative system during the first half-period.

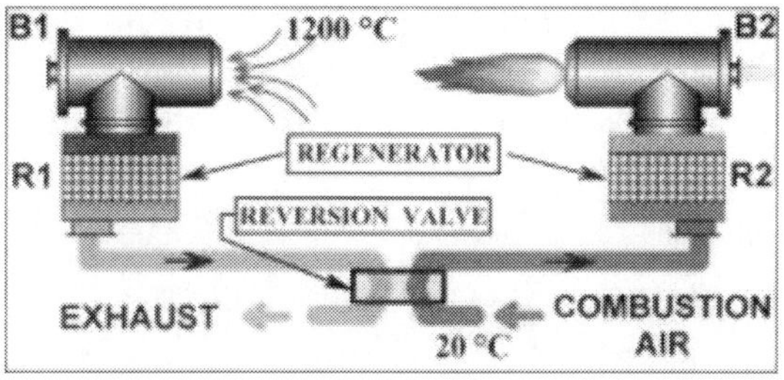

Figure 2. Regenerative system during the second half-period.

3. Assumptions

To analyze the material suitability for the regenerator construction the five materials have been indicated, including two Al_2O_3 and two SiC ceramics types and one Ni-Cr steel. The most important properties of these materials are presented in Table 1. Moreover, density and specific heat are two properties used extensively in the thermodynamic analysis. The product ρc_p commonly termed the volumetric heat capacity, measures the ability of a material to store thermal energy [11]. It is generally recognized that a very good energy storage media are characterized by the value of ρc_p>1 MJ/m^3·K. As may be seen (Table 1) each of considered materials fulfills this relation, so it can be assumed that they might be used to produce the regenerative filling.

Table 1. The properties of materials for regenerator filling [12,13].

Commercial mark or symbol	Type	Density ρ (kg m^{-3})	Thermal conductivity k (Wm^{-1}K^{-1})	Specific heat c_p (J kg^{-1}K^{-1})	Volumetric heat capacity ρc_p (MJ m^{-3}K^{-1})
AluSIK99ZA	Al_2O_3	3800	18	2100	7.98
AluSIK60ZB	Al_2O_3	2300	2	1900	4.37
CarSIK	SiC	3100	50	1700	5.27
CarSIK-NG	SiC	2850	12	1100	3.14
H25N20S2	Steel	7900	19	500	3.95

The honeycomb regenerator filling defined by four construction parameters (Figure 3), such as L – regenerator length, a – internal dimension of the cell, g_w – wall thickness and n – a number of cells in the whole regenerator, has been considered. All these parameters influence the regenerator mass and without g_w they influence the heat transfer area. The regenerator filling was selected for low-power natural gas burner – 30 kW and 6-sectional regenerator design was admitted in considerations.

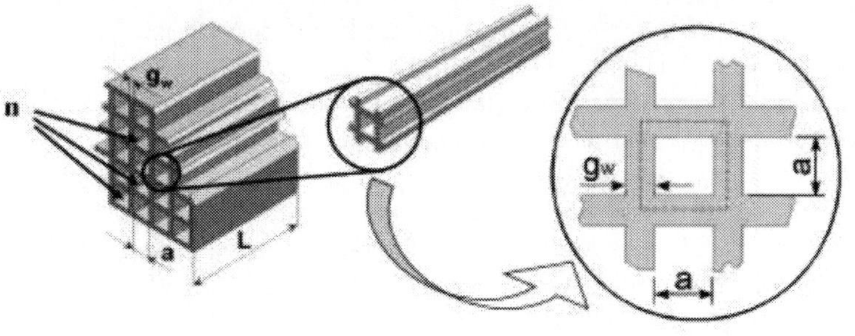

Figure 3. The honeycomb construction parameters.

IOP Conf. Series: Materials Science and Engineering **35** (2012) 012022 doi:10.1088/1757-899X/35/1/012022

Having regard to the possibility of widely available and inexpensive technologies to regenerator produce the following geometric parameters were chosen for the regenerative filling: L = 200 mm (each segment), a = 3 mm, g_w = 2 mm and n = 867. The natural gas, characterized by the calorific value of gaseous fuel of around 34556.8 kJ/m^3, has been chosen to consideration. The fluid mass flows have been calculated according to the power of natural gas burner and for equivalence ratio of 1.05. Thereby the following flows have been obtained: gas mass flow - $6.44 \cdot 10^{-4}$ kg/s, air mass flow - $1.08 \cdot 10^{-2}$ kg/s and exhaust mass flow - $1.15 \cdot 10^{-2}$ kg/s. The inlet fluid temperatures such as an air temperature of 25 °C and the exhaust temperature of 1100 °C have been assumed. The numerical simulation in CFD software has been used as a research method. The numerical calculations have been conducted up to the moment when the semi-steady state of regenerator filling was achieved. The one of the most important value – reversion time – has been adjusted to 15 s.

4. Results analyses

A lot of different kinds of values can be checked in CFD software but the most important comparative value is outflow air temperature in presented research. The used software includes four kinds of averages: area-weighted average, mass-weighted average, facet average and vertex average [14]. However, the facet average has been pointed to calculations.

To compare the heated air temperature the appropriate dynamic characteristics have been prepared according to obtained results. One example of the course of characteristics is presented in Figure 4. It demonstrates as though the nature of regenerator behavior. The two time ranges can be distinguished on the graph – starting phase and pseudo-steady state. Whereas, the most important is the average temperature of preheated air which appears in the pseudo-steady state what is indicated in Figure 4 as well.

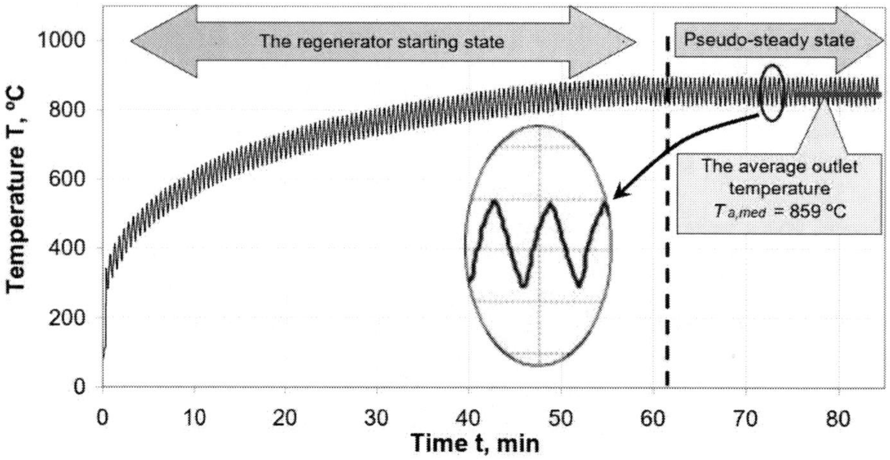

Figure 4. The regenerator time characteristic.

It is also easy to note that temperature (the regenerator filling as well as flowing fluids) changes during the whole period from maximum to minimum because of the regenerator works in an unsteady state. Whereas, when the pseudo-steady state of regenerator is achieved the temperature fluctuation is constant in each of the next periods, as presented in Figure 4 and Figure 5. This course type of the dynamic characteristics is the basis to determine the average temperature ($T_{a,med}$) of the preheated air. However, it should be calculated for whole period in the range from t_1 to t_2 according to formula (1), the graphical interpretation of which is also presented in Figure 5.

$$T_{a,med} = \int_{t_1}^{t_2} T_a(t)dt \cdot (t_2 - t_1)^{-1} \tag{1}$$

where: t_1, t_2 – the beginning and the end of time range for average calculation; $T_a(t)$ – measured preheated air temperature.

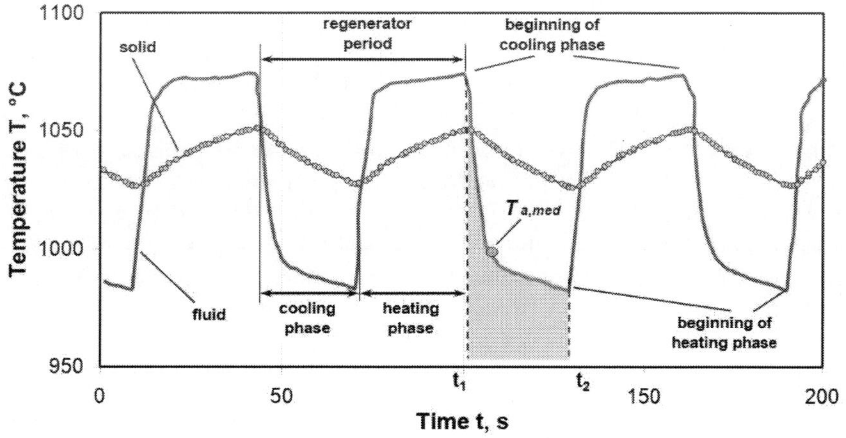

Figure 5. The regenerator pseudo-steady state characteristic $T_a(t)$ with the operation phases indicators and graphical interpretation of the average temperature calculation.

The average temperature of the air preheat calculated from (1) has been taken to the comparison of considered materials. The obtained results have been presented in Figure 6 and Table 2.

Additionally, the time of pseudo-steady state achieving is presented in Table 2, which is also called the starting time. It can be seen that the metal filling is characterized by the one of the shortest starting times compared to the other materials. The explanation for this phenomenon may be such a property termed the thermal diffusivity α, which has units of m^2/s and is defined as the ratio of thermal conductivity to the heat capacity. Materials of large α will respond quickly to changes in their thermal environment, while materials of small α will respond more sluggishly, taking longer to reach a new equilibrium condition.

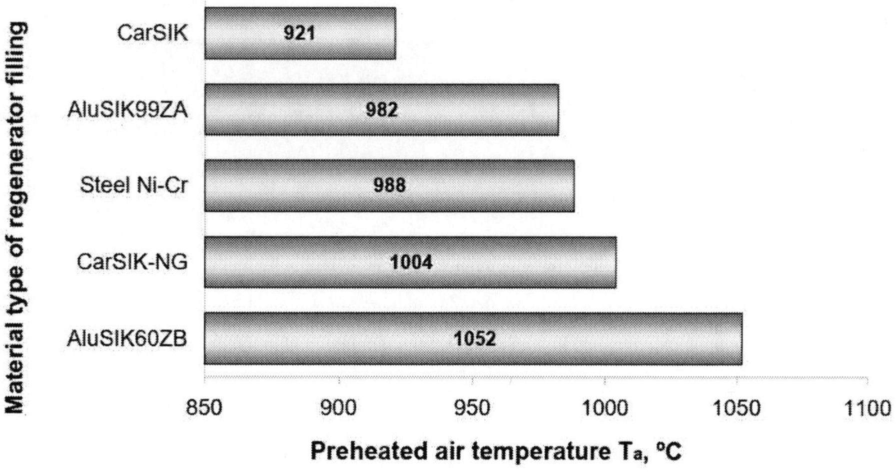

Figure 6. The average of preheated air depending on the regenerator filling material.

The dynamics of the regenerator filling is very important for the quality of steering processes. In this regards, among the considered materials, the metal regenerative filling is a quite good choice. Based on the results comparison it can be stated that other regenerative materials need from 21 to 88 % more time to reach the pseudo-steady state conditions. It means for example, that ceramic filling AluSIK60ZB requires around 19 min more time than metallic filling to achieve the pseudo-steady state, CarSIK requires more than 42 min and AluSIK99ZA needs even more than 82 min. Only the ceramic filling CarSIK-NG is "faster" (by about 20 min) in achieving a pseudo-steady state and it requires around 73 min. The pseudo-steady state time factor has been introduced for better comparison of regenerator dynamics. It is defined as the ratio of time for pseudo-steady state obtaining (for a given material) to the time of the "fastest" material, what is presented in Table 2.

Some kind of inconvenience may include the filling weight. The metal filling is characterized by the biggest mass comparing to the over considered filling materials because of the density, which is the biggest as well. However, for the modern low-power regenerative burners with individual regenerators installed, it is not as so significant because these regenerators are quite small-sized and generally, their absolute weight is not too big in contrast to other, older structures. For considered construction prepared for 30 kW burner, the regenerator mass was from 7 to 11 kg for the ceramic filling and around 22 kg for the metal filling. The mass ratio is presented in Table 2, which defines how many times the metal filling is heavier compared to other materials.

Moreover, to verify the regenerator quality a certain quality indicator has been used as well. The combustion-system efficiency η is that indicator and it has been obtained from the energy balance (according to Figure. 7) which may be expressed as

$$\dot{H}_f + \dot{H}_{air} - \dot{H}_{exh} = \sum \dot{Q} \tag{2}$$

where: $\dot{H}_f$ - fuel enthalpy flow, $\dot{H}_{air}$ - enthalpy flow of the preheated air, $\dot{H}_{exh}$ - enthalpy flow of the exhaust leaving the furnace, $\sum \dot{Q}$ - used or lost heat flow (heat into the charge, heat loss into the ambient via walls and other losses i.e. furnace leaks).

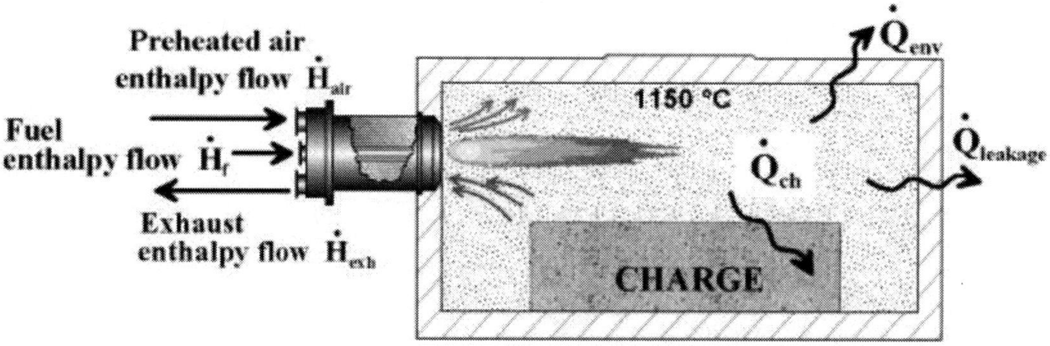

Figure 7. The furnace energy balance with regenerative burners.

Converting (2), it follows (3) where the denominator is defined as combustion-system efficiency η.

$$\dot{H}_f = \sum \dot{Q} \cdot \left(1 + \frac{\dot{H}_{air}}{\dot{H}_f} - \frac{\dot{H}_{exh}}{\dot{H}_f}\right)^{-1} \tag{3}$$

IOP Conf. Series: Materials Science and Engineering **35** (2012) 012022 doi:10.1088/1757-899X/35/1/012022

Based on research results the regenerator quality has been checked according to η. The obtained combustion-system efficiency calculations are also presented in Table 2. The value of 0.865 (for metal filling) is very satisfactory. The efficiency of metal regenerator exceeds the CarSIK and AluSIK99ZA filling efficiency and it is slightly lower than the CarSIK-NG filling. Basically, the obtained result clearly indicates that the high-temperature steel is a good material for the regenerator build, a very high air preheat is reachable and it can be used successfully as a replacement for the regenerative ceramic filling.

Table 2. The summary results of considered materials.

Commercial mark	Thermal diffusivity $\alpha \cdot 10^7$ (m^2s^{-1})	Preheated air temperature $T_{a,med}$ (°C)	Mass ratio m(steel)/m(mat)	Time of pseudo-steady state achieving t_{ps} (min)	Pseudo-steady state time factor t(mat)/t(fast)	Combustion system efficiency η
AluSIK99ZA	22.6	981.8	2.1	175	2.40	0.862
AluSIK60ZB	4.6	1052.2	3.4	112	1.54	0.895
CarSIK	94.9	921.0	2.6	135	1.85	0.834
CarSIK-NG	38.3	1003.6	2.8	73	1.00	0.872
H25N20S2	48.1	987.7	1.0	93	1.27	0.865

The possibility of regenerative burners using for heat furnace has also been considered. The following furnace has been examined: furnace power 4.5 MW; productivity 6 Mg/h; combustion air factor 1.05; exhaust temperature 1100 °C. The metal filling (construction the same as previously discussed) has been used in the regenerators. In the assumptions, changing the old firing system would cause the combustion air temperature increasing from 300 °C up to 987 °C (adopted from research).

As a result of carried out modifications the rate of gas consumption has been reduced from 70.05 to 50.55 m^3/Mg. In this way, the reduction in fuel consumption of 117 m^3/h has been obtained which represents about 27.8%, as well as the furnace efficiency has been increased by 12.2% (from 31.7 to 43.9%). The reduction of fuel consumption causes also the reduction of CO_2 emission. Considering the one month of furnace operation, the fuel saving of $8.7 \cdot 10^4$ m^3 (65 Mg) could be obtained. Assuming an average price of gas as 0.0356 €/kWh [15] the savings of about €29800 could be achieved. Besides, the about 165 Mg of CO_2 will not be emitted into the environment which is an important environmental aspect.

5. Conclusion

From the obtained results can be concluded that there is ability to use the metal materials as the regenerator filling (as a replacement for ceramics) in the heat technologies where the regenerative burners are used. The preheated air temperature is comparable between the metal and ceramic regenerator filling in some cases. For that cases a preheat temperature difference of a few degrees may be insignificant according to the manufacturing costs, which depend on the type of material and complexity technology of resulting from the construction.

The mass of metal filling comparing to ceramic filling should also be neglected as a criterion of material suitability (in the case of individual compact regenerators) because the absolute regenerator weight is not big and more important is the preheated air temperature and manufacturing costs.

Moreover, the metal filling is characterized by the one of the shortest starting times compared to the other materials, as research indicated. Some of the presented materials have been characterized the starting phase even two times longer than metal filling. In connection with that, this attribute plays a very significant role in the steering processes of burners and heat furnaces. Therefore the choice of the "fast" filling is highly desirable and so the metal regenerator filling for this case would be a very good solution.

A certain inconvenience of using the metal filling is the temperature limitation. The metal regenerative filling can be used in heat processes where the exhaust temperature, leaving the combustion chamber, does not exceed 1150 °C.

The use of metal regenerative filling (as a substitute of ceramic filling) characterized by the lower manufacturing costs and satisfactory the combustion air heating, which has a result in consumption fuel reduction and in the limitation of CO_2 emissions, may be an attractive economic aspect and an important environmental aspect as well.

References

[1] Hasegawa T 2000 Environmentally-compatible regenerative combustion heating system *The 2nd Int. Seminar on High Temperature Combustion (Stockholm, Sweden, 17-18 January 2000)*

[2] Weber R 2002 New Combustion Technology for Reheating Furnaces *Challenges in Reheating Furnaces (London, UK, 28-29 October 2002)* p 163.

[3] Teufert J Domagala J 2009 Regenerative burner systems for batch furnaces in the steel industry *Gaswärme International* **58** 51

[4] Tomeczek J Wnek M 2005 Regenerative gas burners for high-temperature heating furnaces *Hutnik* **9** 461-465

[5] Wnek M 2011 The compact heat regenerator as a solution of fuel consumption reduction and CO_2 emission reduction *Energy Transformations in Industry - 12th International Scientific Conference (Low Tatras - Jasna pod Chopkom, Slovakia, 21-23 September 2011)* (Kosice: Acta Metallurgica Slovaca Conference vol.2 no.1) pp 243-248

[6] Wünning JG 2006 Small Capacity Regenerative Burners *2007 AFRC - JFRC International Symp. (Hawaii, USA, 16-18 October 2006)*

[7] http://www.namfg.com *TwinBed II Regenerative Burners* [15.03.2012]

[8] Tomeczek J Puszer A Wnęk M 2004 Possibility of fuel consumption reduction in the continuous heating furnaces *Hutnik* **4** 178-181.

[9] Manatura K Tangtrakul M 2010 A Study of Specific Energy Consumption in Reheating Furnace Using Regenerative Burners Combined with Recuperator *Silpakorn U Science & Tech J* **4** 7-13.

[10] http://www.bloomeng.com/regenerative.html [15.03.2012]

[11] Incropera F P De Witt D P 2002 *Fundamentals of Heat and Mass Transfer* (USA: John Wiley and Sons)

[12] Schunk Ingenieurkeramik GmbH 02.94/1000 Ex

[13] Certificate UGITECH Groupe Arcelor France for IMS Stalserwis [12-08-2004]

[14] Fluent User's Guide.

[15] http://www.energy.eu [14.03.2012]

AUTHOR INDEX

Institute of Physics
Dirac House, Temple Back
Bristol BS1 6BE UK

CURRAN ASSOCIATES INC
proceedings.com